Game Theory - Computational Aspects and Applications

Edited by Tibor Guzsvinecz and Judit Szűcs

Published in London, United Kingdom

Game Theory - Computational Aspects and Applications
http://dx.doi.org/10.5772/intechopen.1001635
Edited by Tibor Guzsvinecz and Judit Szűcs

Contributors
Adrian Northey, Ali Moridi, Dandan Wang, David Lambert, Eyup Akcetin, Hao Lei, Ian M. Robinson, Jianlin Zhu, Jingjing Chen, Judit Szűcs, Mark Flanagan, Mohammad Ali Tolouei Virani, Reza Javidi Sabbaghian, Tibor Guzsvinecz, Tomas By, Trevor C. Lipscombe, Yanzhao Bi, Youmei Zhou

First published in London, United Kingdom, 2025 by IntechOpen
IntechOpen is the global imprint of INTECHOPEN LIMITED, registered in England and Wales, registration number: 11086078, 167-169 Great Portland Street, London, W1W 5PF, United Kingdom

For EU product safety concerns: IN TECH d.o.o., Prolaz Marije Krucifikse Kozulić 3, 51000 Rijeka, Croatia, info@intechopen.com or visit our website at intechopen.com.

British Library Cataloguing-in-Publication Data
A catalogue record for this book is available from the British Library

Game Theory - Computational Aspects and Applications
Edited by Tibor Guzsvinecz and Judit Szűcs
p. cm.
Print ISBN 978-0-85014-832-9
Online ISBN 978-0-85014-831-2
eBook (PDF) ISBN 978-0-85014-833-6

If disposing of this product, please recycle the paper responsibly.

Meet the editors

Dr. Tibor Guzsvinecz obtained his Ph. D. in Information Science from the University of Pannonia in Hungary in 2021. His research areas include data science, gamification, human-computer interaction, spatial ability, video games, and virtual reality. He has 108 scientific publications and has edited four book projects. He has a Hirsch index of 12. He received eight scholarships, one section chair award from the Institute of Electrical and Electronics Engineers (IEEE), and a best publication award at a conference. He is a member of the international organizing committee of the IEEE International Conference on Cognitive Infocommunications and the IEEE International Conference on Cognitive Aspects of Virtual Reality. He is also a member of two laboratories at the University of Pannonia in Hungary. He has been an IEEE member since 2023.

Dr. Judit Szűcs obtained her Ph. D. in Computer Science at the University of Szeged in Hungary in 2021. Her main interests include binary tomography, image processing, and computer vision. She has 55 scientific publications, including 12 journal papers and 19 studies in various conference proceedings. She has received four scholarships, two best poster awards and the Tamás Roska Scientific Presentation award. As a Ph. D. student, she also received the Bronze Award from the IEEE Hungary Section on a student paper contest. She has been an IEEE member since 2023.

Contents

Preface

In an era characterized by rapid advancements in science, technology, and interconnected systems, a need arises for sophisticated tools to analyze, model, and resolve complex problems. To address such challenges, game theory has emerged as a branch of mathematics and economics. Game theory provides a better understanding of the decision-making processes and conflict resolution by systematically examining strategic interactions between individuals, organizations, or even systems. This book serves as a collection of contributions that exemplify the diverse applications of game theory across multiple domains, offering readers a comprehensive understanding of its theoretical foundations and real-world applications.

The chapters in this book are organized into four sections, each highlighting a distinct aspect of game theory's versatility.

The first section, "Game Theory in Computational and Analytical Frameworks", shows computational methodologies that enable precise and replicable modeling of games. Tomas By's chapter on the G-Code Game Play Formalism discusses the importance of game definition languages for simplifying game development and ensuring the scientific rigor of experiments. Complementing this, Mark Flanagan and his co-authors explore Lanchester's heterogeneous equations in their analysis of mixed-for competitions, illustrating how game theory can predict outcomes in wargame scenarios with ternary mixtures. Together, these chapters underscore how computational tools amplify the reach and depth of game-theoretical analyses.

The second section, "Applications of Game Theory in Strategic Decision-Making", shows how game theory is used in not only traditional fields to inform strategic planning. Youmei Zhou and colleagues demonstrate how evolutionary game theory can facilitate decision-making in urban renewal and rural tourism planning, providing an understanding of stakeholder dynamics and evidence-based policy development. On a different note, Tibor Guzsvinecz and Judit Szűcs apply game theory to the Dark Souls video game series, illustrating how strategic concepts like risk management and interactive learning can portray player behavior in high-stakes, decision-intensive environments. These chapters show the adaptability of game theory in analyzing diverse decision-making contexts, from urban systems to digital landscapes.

The third section, "Game Theory in Resource Management and Allocation", examines its application in resolving conflicts over shared resources, such as water. Ali Moridi's chapter focuses on managing water quality disputes and shows how game theory models stakeholders interact to suggest realistic and innovative solutions. Similarly, Reza Javidi Sabbaghian and Mohammad Ali Tolouei Virani examine game-theoretical bankruptcy methods for allocating shared water resources equitably. Their work shows how cooperative game theory provides alternatives to conventional optimization models, mainly in scenarios of scarcity and competing interests. This section illustrates the potential of game theory to foster collaboration and sustainable solutions in critical resource management contexts.

The final section, "Integrative and Interdisciplinary Approaches", expands the boundaries of game theory by linking it with other scientific and analytical frameworks. Eyup Akcetin's chapter details how to integrate systems thinking, decision intelligence, and swarm intelligence to optimize business performance. This chapter demonstrates how businesses can achieve strategic foresight and long-term success by combining game theory with interdisciplinary tools and real-time data analytics. This section shows the transformative potential of combining game theory with other methodologies to solve complex, multi-dimensional problems.

Collectively, this book offers an in-depth exploration of game theory as both a theoretical discipline and a practical tool. From computational innovations and strategic planning to resource management and interdisciplinary integrations, the chapters provide novel ideas that appeal to academics, practitioners, and policymakers alike. This book shows the versatility of game theory and its ability to adapt to the unique challenges of different fields. We hope this book inspires readers to explore new applications of game theory in their respective areas and fosters further innovation in this evolving field.

We encourage you to consider the theoretical results presented here and their practical implications in the real world. Whether you are a researcher, a strategist, or simply curious about the decision-making process, this book is designed to provoke thought and encourage further exploration of game theory's boundless possibilities.

Tibor Guzsvinecz and Judit Szűcs
Department of Information Technology and its Applications,
University of Pannonia,
Zalaegerszeg, Hungary

Section 1

Game Theory in Computational and Analytical Frameworks

Chapter 1

The G-Code Game-Play Formalism

Tomas By

Abstract

The main purposes of a game definition language are: to simplify the task of implementing the game, compared to coding it directly in a programming language; to define a standard representation, ensuring that everybody uses the exact same definition of the game, thereby enabling scientific methods of operation such as repeatability of experiments and clear communication of results; and finally to serve as a long-term repository, releasing the game from dependence on computer hardware and operating systems. The major problems that the human writer of the game definition faces are: how to turn the game rules into statements in the formalism, and how to ensure that the game definition as a whole is complete and correct. This seems to suggest that a helpful design of a game definition language is one where the game definition closely follows a description of the game written for human consumption; where standard devices, such as tables and gridded mapboards, can be encoded easily and succinctly; and where the game definition is clear and compact.

Keywords: game-play, formalisation, little language, formal language, game logic

1. Introduction

The rules of computer games are normally not available to the players in full detail. From the player's point-of-view, this is not generally a problem, and may even add to the challenge and enjoyment of playing the game. Technically, the game rules, in their final form, often only exist as computer source code and are therefore not easily comprehensible for people who are not experts in the specific computer language and other related technologies, even if they can access the code.

It seems clear that any deeper scientific study of a game requires that the game rules be known and also available computationally, which means formally, not expressed in human language. This latter issue also applies to manual games that are not available as software. Before meaningful computational experiments and studies can be performed on those games, the rules must be expressed in a computational formalism. A solution to this problem, which is a formal language for expressing the game-play, including all the rules and also the relevant properties of the game board and the playing pieces, was proposed in [1] and has been further developed, in a leisurely fashion, since. There have not been any significant changes, only minor additions.

This chapter explains how this formalism works, using three complete examples: standard *Chess* (Appendix A) and the board wargames *Kassala* (Appendix B; [2, chap 4]) & *Drive on Metz* (Appendix C; [3, pp. 174–89]). Both of the latter are pedagogical example games, with smaller game boards and simpler rules than what is customary in these types of games. Although the G-code formalism is mainly focussed on wargames (cf. [4, 5]), it is designed to be completely general and handle any board game (cf. [6, 7]). Computer games that rely on execution speed and fast network communication are not expected to be practical to formalise using G-code.[1]

After the description of the formalism in the next section, follow brief comments on future work (Section 3), comparisons to other similar proposals in the literature (Section 4), and general conclusions (Section 5).

2. G-code

A game definition in G-code encodes the rules, including all possible actions and the victory conditions, and all the properties of the game board and playing pieces that are relevant for the rule logic. Purely æsthetic properties are not in G-code (but there is a parallel encoding system, not described here, for the graphics). Complete examples of G-code game formalisations are given in Appendix.[2]

At the top level, a G-code formalisation contains sections and individual statements. All these can be in any order. Inside some of the sections, the order of statements matter, as will be described later on. The types of sections are the following.

`sop`	The 'sequence of play'. Exactly one of these is required.
`map`	One single map board. There can be several of these sections.
`table`	Tabular data, as is quite common in games.
`action`	Defines a player's action. Otherwise, same as the procedure.
`procedure`	Similar to namesakes in programming languages, sequences
`function}`	of operations collected under one name.

As is common in computer programming, there can be many ways to achieve the same effect. The tables, for example, are mainly for convenience, so that the G-code formalisation is as close as possible to the printed game rules. Technically, it is perfectly possible to implement the same logic as a function containing a series of conditional statements.

The individual statements at the top level are of these types:

`terrain`	Defines one map terrain type.[3]
`type`	Defines a discriminated type (i.e. values not ordered).
`enum`	Defines an enumerated type (i.e. values ordered).
`define`	Defines an 'object' type (attribute-value structure).
`object`	Defines a propertyless object.[4]
`relation`	Defines a relation type.[5]
`variable`	Defines one variable.

---

1 But who knows. If computer technology continues to develop in the way it has over recent decades, then current 'first person shooter' multi-player games may at some point be formalisable as well.

2 The well-known board game *Chess* in Appendix A, and the simplified wargames *Kassala* ([2], chap 4) in Appendix B & *Drive on Metz* ([3], pp. 174–89) in Appendix C. The very simple wargame *Strike Force One* is shown in Ref [1].

3 At the top level, it applies to all maps. It can also be inside the map definition.

4 Used for things like off-map areas and abstract production resources.

5 All the examples use a relation '`location`' for placing units on the map.

The 'terrain' statement can be used also inside a map section (c.f. footnote 3) and the 'variable' definition can occur inside action/procedure/function definitions. The other individual statement types must be at the top level. The built-in types are integers, floating-point numbers, and booleans.

2.1 Sequence of play and victory conditions

Like the other sections, the 'sop' section can be anywhere in the G-code file, but it is the starting point of the dynamic behaviour, and typically contains a loop and some calculations that determine who won or lost. In *Chess* (Appendix A), each step contains one single action, the players alternating, and the game ends when wither king is no longer on the board.

```
begin sop
   player := white;
   begin loop
      dialogue one{ king,queen,knight,bishop,castle,pmove,pcap,
epass,pprom} ;
      begin if not ( location ( WhKg ) & location ( BlKg ) ) then
         exit;
      end if
      player := enemy(player);
   end loop
   winner := player;
end sop
```

The wargame *Kassala* (Appendix B) has a more complex arrangement where each turn has four phases, and each of those involves a number of actions depending on conditions on the board.

```
begin sop
   variable winner : winner;
   variable ck, cu : set(unit);
   begin for[ i:int, 1..10]
      begin foreach unit u : u.mf:=u.mfmax; u.haf:=false; end foreach
      dialogue many{ move(moslem)} { stacking} ;
      dialogue many{ combat(moslem)} { mustfight} ;
      dialogue many{ move(christian)} { stacking} ;
      dialogue many{ combat(christian)} { mustfight} ;
   end for
   ck = christians_in(kassala);
   cu = christians_in(udaka);
   begin if ( empty(ck) & empty(cu) ) then
      winner := moslem;
   else if ( not empty(ck) & not empty(cu) ) then
      winner := christian;
   else
      winner := draw;
   end if
end sop
```

The '`dialogue`' statement indicates that input from the player is required, either one single choice or a sequence of selections until some ending condition is fulfilled.[6]

2.2 Game board

In *Chess* (Appendix A), the colours on the board have no function in the game, so are not indicated in G-code, and the only information that needs to be given is that it is an 8 × 8 square board. The numbering is indicated using the format described in [8].

```
begin map board
type square 8 8;
coordinates "A.+90.Nd";
end map
```

The wargames *Kassala* (Appendix B) and *Drive on Metz* (Appendix C) have hexagonal boards of roughly the same size as *Chess*, but they also have terrain. The entire map definition in *Kassala* is shown in **Figure 1b**, and *Drive on Metz* contains the following:

```
type hexagonal 9 11;
coordinates "0N.-90.0Noi";
```

The game *Kassala* as given in [2, chap 4] does not use any numbering of the locations, but the one used here is standard in wargames. *Drive on Metz* has a number printed on the map board (**Figure 2**).

2.3 Playing pieces

The usual way to define the playing pieces and the positioning on the board is to use 'objects'[7] and a 'relation'. In a game like *Chess*, it would be possible to instead let the piece be an attribute of the location, or vice versa, but using a relation is a more general solution, as it easily handles things like multiple pieces in one location or special off-map locations, as is common in wargames.

In *Chess* (Appendix A), there are six types of pieces, and they are located in the squares on the board.

```
define piece{ colour:colour, ptype:ptype} ;
type ptype = king | queen | bishop | castle | knight | pawn;
relation location : piece <-> square;
```

The individual pieces are then defined, and their starting locations defined as a relation.

```
piece WhKg{ colour=white, ptype=king} ;

location : WhKg <-> board.A5;
```

The white king starts at location A5.

---

[6] In *Chess*, pmove = pawn move; pcap = pawn capture; epass = en passant; & pprom = pawn promotion.

[7] Not the same as in object-oriented programming, but simply typed attribute-value records.

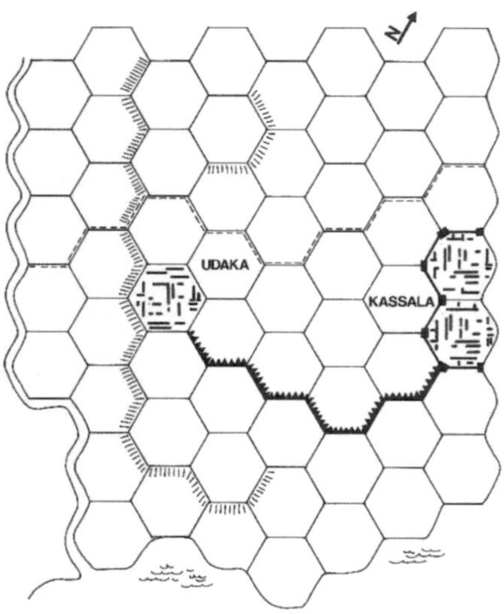

(a) Complete map

```
begin map kassala
type hexagonal 8 8;
coordinates 0Nou.+90.0N;
terrain loc clear; terrain loc kassala;
terrain loc udaka;
terrain side[directed] wadi;
terrain side[directed] trench;
clear: 0104, 0105, 0106, 0107, 0108, 0201,
0202, 0203, 0204, 0205, 0206, 0207, 0208,
0302, 0303, 0304, 0306, 0307, 0308, 0401,
0402, 0403, 0404, 0405, 0406, 0407, 0408,
0501, 0502, 0503, 0504, 0505, 0506, 0507,
0508, 0601, 0602, 0603, 0604, 0605, 0606,
0607, 0608, 0702, 0703, 0704, 0705, 0706,
0707, 0708, 0802, 0803, 0806, 0807, 0808;
kassala: 0804, 0805;
udaka: 0305;
wadi: 0208>0308, 0207>0208, 0207>0307,
0206>0307, 0206>0306, 0205>0306, 0205>0305,
0204>0305, 0204>0304, 0203>0304, 0203>0303,
0202>0303, 0302>0303, 0302>0402, 0401>0402,
0302>0402, 0406>0407, 0507>0407, 0508>0407;
trench: 0404>0304, 0404>0403, 0504>0403,
0504>0503, 0603>0503, 0603>0602,
0603>0703, 0704>0703, 0704>0803;
end map
```

(b) G-code

Figure 1.
Kassala [2].

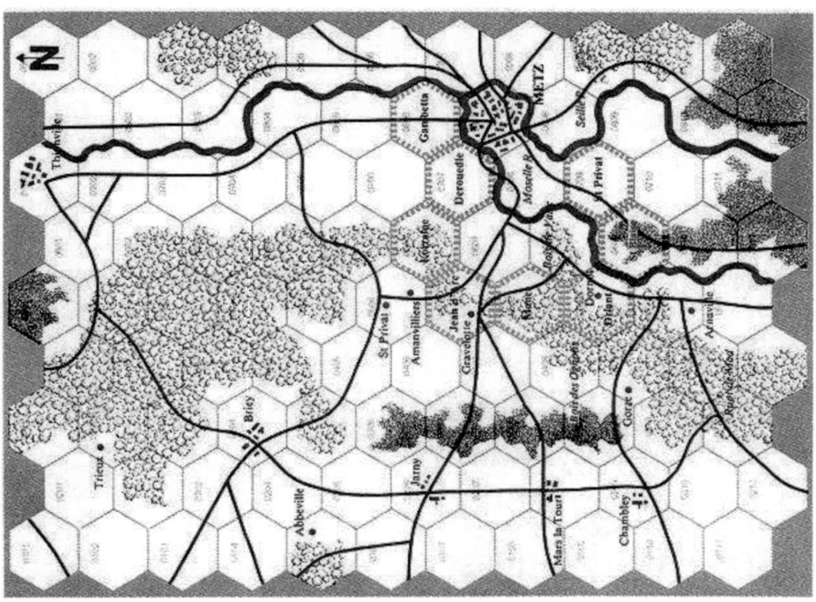

Figure 2.
Drive on Metz [3].

In *Kassala* (Appendix B), the units have more attributes. The value 'mf' is the current amount of movement factors, which decreases for each move the unit makes, and 'mfmax' is the amount at the start of each turn. The attribute 'haf' indicates whether the unit has been involved in fighting, which is not allowed to do more than once per turn.

```
define unit{ player:player, type:type, str:int, mf:int,
mfmax:int, haf:bool} ;
type type = infantry | cavalry | cannon;
relation location : unit <-> hex;
```

Then the actual units and their starting locations are defined, same as in the *Chess* example.

```
unit GE1{ player=christian, type=infantry, str=4, mfmax=2} ;
location : GE1 <-> kassala.0804;
```

In the game *Drive on Metz* (Appendix C), not all units are on the map at the start of the game, and when new units arrive, they enter the side of the board, and they can be kept off the board as long as the player wishes. To handle this, a force pool is defined as an object and defined as a possible location.

```
define unit{ player:player, str:int, mp:int,
mpmax:int, haf:bool} ;
object pool;
relation location : unit <-> ( hex | pool );
```

Otherwise, it works the same as in the other examples.

```
unit "17SS/38SS"{ player=German, str=2, mpmax=8} ;
location : "17SS/38SS" <-> metz.0709;
```

The German 38th ᛋᛋ armoured infantry regiment starts at the location 0709.

2.4 Actions

The 'actions' are those things that the player can do in the game. In the *Chess* example (Appendix A), there is one for each of the different types of pieces. It is not necessary to design this way; there could be just one single 'move' action and then a conditional on the type.

```
begin action pawn ( ptype p, square from, square to )
   p.ptype = pawn;
   location ( p <-> board.from );
   not board.to.ptype = p;
   d := distance(from,to);
   ( p.ptype = white & d = { 1,_} ) or ( p.ptype = black & d = { -1,_} );
   begin if ( location ( x <-> board.from ) & x.ptype \= p ) then
      retract ( location ( x <-> board.from ) );
   end if
   retract ( location ( p <-> board.from ) );
   assert ( location ( p <-> board.to ) );
end action
```

An action definition starts with conditions that have to be fulfilled, in this case that the piece is a pawn, that the destination is not already occupied by another of the same player's pieces, and that the move is one step forward.

As can be expected, the actions in wargames are often more complicated. The combat actions in *Kassala* (Appendix B), for example, have a number of loops and local variables.

```
1 begin action combat ( player p, set(unit) as, set(unit) ts )
2     begin foreach unit a in as : a.haf = false; end foreach
3     begin foreach unit t in ts : u.haf = false; end foreach
4     begin foreach unit aa in as :
5         begin foreach unit tt in ts :
6             variable al, tl : hex;
7             is_instance ( location ( aa <-> al ) );
8             is_instance ( location ( tt <-> tl ) );
9             distance ( kassala, al, tl ) = 1;
10          end foreach
11     end foreach
12     variable hs : set(hex) := empty;
13     begin foreach unit tx in ts :
14         variable x : hex;
15         is_instance ( location ( x <-> tx ) );
16         insert(x,hs);
17     end foreach
18     begin case crt[ basic_ratio (as,ts)][ random(1,6)]
19         DE : eliminate(ts); advance(as,hs);
20         AE : eliminate(as);
21         EX : exchange(as,as,ts);
22              begin if not empty(as) then
23                  advance(as,hs);
24              end if
25         C : begin end
26     end case
27     begin foreach unit u in as : u.haf:=true; end foreach
28     begin foreach unit ttt in ts : ttt.haf:=true; end foreach
29 end action
```

The first two lines (2 & 3) are the condition that all the units that are presently attacking or defending cannot already have fought in the same turn. The only difference between these lines and the bottom two lines (27 & 28) is the comparison (=) instead of assignment (:=). Then, the nested loops in lines 4–11, check the condition that all attacking units be adjacent to all defending units (and vice versa). The code in

SIMPLIFIED COMBAT ODDS								
DIE ROLL	1–4 or less	1–3	1–2	1–1	2–1	3–1	4–1	5–1 or more
1	C	EX	EX	C	DE	DE	DE	DE
2	C	C	EX	EX	C	DE	DE	DE
3	AE	C	C	EX	EX	C	DE	DE
4	AE	AE	C	C	EX	EX	C	DE
5	AE	AE	AE	C	C	EX	EX	C
6	AE	AE	AE	AE	AE	C	EX	EX

(a) Original table

```
begin table crt [ratio[1/4|5/1] x int[1-6] => cmbt_res]
  "1-4<" "1-3" "1-2" "1-1" "2-1" "3-1" "4-1" "5-1>"
1 C      EX    EX    C     DE    DE    DE    DE
2 C      C     EX    EX    C     DE    DE    DE
3 AE     C     C     EX    EX    C     DE    DE
4 AE     AE    C     C     EX    EX    C     DE
5 AE     AE    AE    C     C     EX    EX    C
6 AE     AE    AE    AE    AE    C     EX    EX
end table
```

(b) G-code

Figure 3.
Kassala combat results table [2, p. 55].

lines 12–17 just collects all the locations of the defending units (`hs`), so that they are available in case the combat result includes 'advance'. Lines 18–26 are possible outcomes, depending on the die roll. The '`crt`' in line 18 is the name of the combat results table (**Figure 3b**). As can be seen in **Figure 3**, the format in G-code is very similar to the printed format in the rules. This is important, as it makes it easier to create the game formalisation, compared to having to code the table information as a set of conditionals, for example, and it also makes it much easier to verify that the formalisation expresses the correct rules of the game.

3. Current status and expected future developments

Additions in the current version, beyond what is described in [1], include multiple mapboards and a more general numbering system [8]. Planned additions are combinations of rotated maps, which is not uncommon; units that are larger than one hexagon, which is not very common but reasonably easy to do; and irregular areas, used for things like burst weapon effects in tactical games.

Things that are being designed and are planned to be included in the next version are unit facing and fog-of-war. These are not complicated technically, but somewhat less obvious how to manage unobtrusively in the formalism. It is also possible that the range of possible forms of the '`dialogue`' statement may need to be extended.

The only presently known mechanism in current board games that will not be included in G-code in the foreseeable future is when graphics on the map influence play. An example is shown in **Figure 4**, from the game *firepower*. The line-of-sight from the rifleman just outside the house in the lower left to the enemy across the road is broken by the corner of the house, preventing him from firing. Line-of-sight is traced from hexagon centre to hexagon centre. But the exact location of the house walls is not computationally available, and even if it is easy to scan an image, it is not clear how to appropriately represent (for example) the house wall location data in a general way in the game.

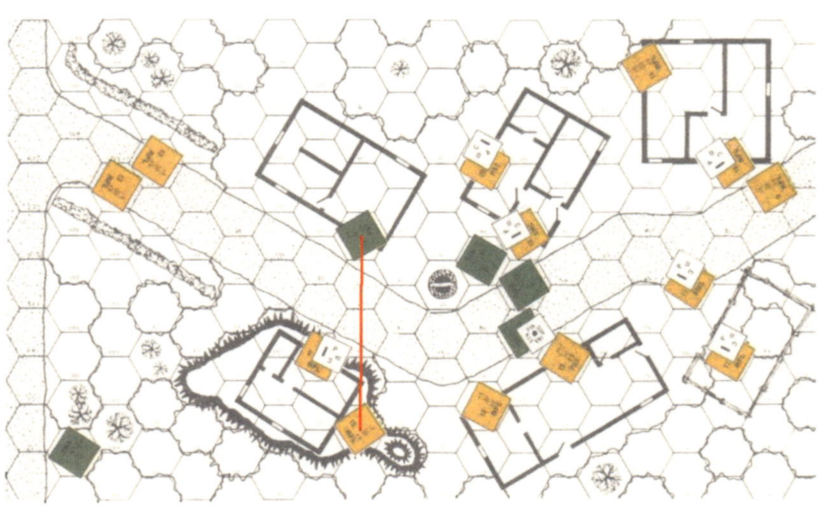

Figure 4.
Firepower line-of-sight broken by building.

4. Comparison with other similar systems

Various systems exist for computerising multiple games within the same framework. The idea is not new (cf. [9]). There were a couple of general wargaming systems,[8] now obsolete because of advances in computer graphics and developments in computer operating systems. A current product is *Zillions of games*.[9] More research-oriented work includes RAND-ABEL [10–12], Metagame [13–15], Multigame [16, 17], GDL [18], an 'Engine' [19] and Casanova [20, 21]. Most of these are limited in various ways. Kaiser ([18], p. 21) reports that *Zillions of games* does not allow more than one piece in the same position, which is quite common in wargames. RAND-ABEL is primarily meant for implementing game-playing agents rather than formalising the whole game. Both Metagame and Multigame assume square grid, perfect information and no randomness. Casanova may be general, but is apparently focussed on the graphics and real-time play, not the abstract game logic. From a more technical point-of-view, some of these are extensions of programming languages,[10] and some[11] use Lisp syntax, but seem to be independent formalisms. *Zillions of games* mixes graphical information (e.g., image file names) with the game logic.

From the point-of-view of the game designer,[12] Multigame and the 'Engine' seem reasonably clear and simple. Metagame, *Zillions of games*, and GDL are very logic-like in that information is spread out over a large number of simple statements rather than concentrated in a compact expression, such as a table or a list with a minimal amount of syntactic decoration.[13] Another aspect of the problem of writing down the game rules, besides the expressiveness and compactness of the formalism, is that using an extension of a full programming language means that all the rest of that language is also available, which makes any automatic processing of the formalised game harder. A further factor, perhaps not as important, is that using the syntax of an existing language may invite incorrect assumptions about how things work. G-code uses an Algol-like syntax, which is not common in modern programming languages. It could be that using an entirely new made-up syntax would be even better, but there is probably a fine line between achieving that and incomprehensibility.

Classic board games such as Chess and Go are stable, and have simple rules. In wargames, on the other hand, the fact that they simulate, to some extent, reality, means that there is an intrinsic struggle between 'playability' and 'realism'. Making the game more realistic typically means making the rules more complicated, and there is a risk of seemingly absurd behaviour.[14] Having a formalism that makes it easy to read and understand the rules, and to make changes, seems more important the more complicated the game gets.

---

[8] SSI *Wargame Construction Set & HPS Aide-de-camp*.

[9] www.zillions-of-games.com

[10] RAND-ABEL: C; 'Engine:' Python; Casanova: f#.

[11] GDL and *Zillions of games* at least.

[12] Or the game Converter, that is, the person who converts an existing board game into G-code, which is not necessarily a mechanical operation.

[13] Somewhat similarly, Silver et al. [22] encode the rules of the games (Chess & Shogi) in a very technical, system-specific way, not easily available for inspection and verification.

[14] The best example may be 'Panzerbush'. In the game *Panzerblitz*, rules for spotting before being able to fire meant that units could make short moves over open terrain but always stopping in woods, without the opponent being able to shoot at them. Cf. [23].

5. Conclusions

Games are a limited, well-defined domain, suitable for abstracting away from programming details through a game definition formalism. This formalism needs, of course, to be expressive enough to handle all procedures and mechanisms in some large class of games. It is helpful if the expressions in the game definition language are close to the way things are presented in rules written in human language, as this makes it easier to write the formal definitions and to later verify that they are correct. Tables and the map grid coordinates are perhaps the most obvious examples.

Besides making the job of implementing the game much less work than programming everything from scratch, game definition formalism provides a standard, precise representation of the rules of the game. Given that the formal version is agreed on, there can be no ambiguities in how the rules apply in specific situations, which is not unheard of in human language game rules. For scientific methods of study, such as repeatable experiments and clear communication of results, it is important that everybody uses the exact same definition of the game rules. Finally, a formal definition serves as a long-term repository, ensuring that the game is still available many years later, long after the computer hardware and operating systems have changed beyond recognition.

Appendix A

The well-known board game *Chess.*

```
begin map board
type square 8 8;
coordinates "A.+9 0.Nd";
end map

define piece{ colour:colour, ptype:ptype};
type colour = white | black;
type ptype = king | queen | bishop | castle | knight | pawn;
relation location : piece <—> square;
piece WhKg{ colour=white, ptype=king}; location : WhKg <->
board.A5;
piece WhQn{ colour=white, ptype=queen}; location : WhQn <->
board.A4;
piece WhBp1{ colour=white, ptype=bishop}; location : WhBp1 <->
board.A3;
piece WhBp2{ colour=white, ptype=bishop}; location : WhBp2 <->
board.A6;
piece WhKn1{ colour=white, ptype=knight}; location : WhKn1 <->
board.A2;
piece WhKn2{ colour=white, ptype=knight}; location : WhKn2 <->
board.A7;
piece WhCt1{ colour=white, ptype= castle}; location : WhCt1 <->
board.A1;
```

```
piece WhCt2{ colour=white, ptype= castle} ; location : WhCt2 <->
board.A8;
piece WhPw1{ colour=white, ptype=pawn} ; location : WhPw1 <->
board.B1;
piece WhPw2{ colour=white, ptype=pawn} ; location : WhPw2 <->
board.B2;
piece WhPw3{ colour=white, ptype=pawn} ; location : WhPw3 <->
board.B3;
piece WhPw4{ colour=white, ptype=pawn} ; location : WhPw4 <->
board.B4;
piece WhPw5{ colour=white, ptype=pawn} ; location : WhPw5 <->
board.B5;
piece WhPw6{ colour=white, ptype=pawn} ; location : WhPw6 <->
board.B6;
piece WhPw7{ colour=white, ptype=pawn} ; location : WhPw7 <->
board.B7;
piece WhPw8{ colour=white, ptype=pawn} ; location : WhPw8 <->
board.B8;
piece BlKg{ colour=black, ptype=king} ; location : BlKg <->
board.H5;
piece BlQn{ colour=black, ptype=queen} ; location : BlQn <->
board.H4;
piece BlBp1{ colour=black, ptype=bishop} ; location : BlBp1 <->
board.H3;
piece BlBp2{ colour=black, ptype=bishop} ; location : BlBp2 <->
board.H6;
piece BlKn1{ colour=black, ptype=knight} ; location : BlKn1 <->
board.H2;
piece BlKn2{ colour=black, ptype=knight} ; location : BlKn2 <->
board.H7;
piece BlCt1{ colour=black, ptype= castle} ; location : BlCt1 <->
board.H1;
piece BlCt2{ colour=black, ptype= castle} ; location : BlCt2 <->
board.H8;
piece BlPw1{ colour=black, ptype=pawn} ; location : BlPw1 <->
board.G1;
piece BlPw2{ colour=black, ptype=pawn} ; location : BlPw2 <->
board.G2;
piece BlPw3{ colour=black, ptype=pawn} ; location : BlPw3 <->
board.G3;
piece BlPw4{ colour=black, ptype=pawn} ; location : BlPw4 <->
board.G4;
piece BlPw5{ colour=black, ptype=pawn} ; location : BlPw5 <->
board.G5;
piece BlPw6{ colour=black, ptype=pawn} ; location : BlPw6 <->
board.G6;
piece BlPw7{ colour=black, ptype=pawn} ; location : BlPw7 <->
board.G7;
piece BlPw8{ colour=black, ptype=pawn} ; location : BlPw8 <->
board.G8;
```

```
variable player : colour;
variable winner : colour;

begin sop
   player : = white;
   begin loop
      dialogue one{ king,queen,knight,bishop,castle,pmove,pcap,
      epass,pprom} ;
      begin if not ( location ( WhKg ) & location ( BlKg ) ) then
         exit;
      end if
      player : = enemy(player);
   end loop
   winner := player;
end sop
begin action king ( ptype p, square from, square to )
   p.type = king;
   location ( p <-> board.from );
   not board.to.type = p.type;
   d := distance(from,to); d ={ 1,_} or d ={ _,1} ;
   begin if ( location ( x <—> board.from ) & x.type \= p.type )
   then
      retract ( location ( x <-> board.from ) );
   end if
   retract ( location ( p <-> board.from ) );
   assert ( location ( p <-> board.to ) );
end action

begin action pmove ( ptype p, square from, square to )
   p.type = pawn;
   location ( p <-> board.from );
   not board.to.type = p.type;
   d : = distance ( from, to );
   ( p.type = white & d ={ 1,0} ) or ( p.type = black & d ={ -1,0} );
   retract ( location ( p <-> board.from ) );
   assert ( location ( p <-> board.to ) );
end action

begin action pcap ( ptype p, square from, square to )
   p.type = pawn;
   location ( p <-> board.from );
   not board.to.type = p.type;
   d : = distance (from,to);
   ( p.type = white & ( d ={ 1,-1} or d ={ 1 ,1} ) )
      or ( p.type = black & ( d ={ -1,-1} or d ={ -1,1} ) );
   begin if ( location ( x <-> board.from ) & x.type \= p.type ) then
      retract ( location ( x <-> board.from ) );
   end if
   retract ( location ( p <-> board.from ) );
   assert ( location ( p <-> board.to ) );
```

```
end action

begin action epass ( ptype p, square from, square to )
   p.type = pawn;
   location ( p <-> board.from );
   ???
   retract ( location ( x <-> board.from ) );
   retract ( location ( p <-> board.from ) );
   assert ( location ( p <-> board.to ) );
end action

begin action pprom ( ptype p )
   p.type = pawn;
   location ( p <-> board.from );
   ???
   retract ( location ( p <-> board.from ) );
   assert ( location ( p <-> board.to ) );
end action

begin action knight ( ptype p, square from, square to )
   p.type = knight;
   location ( p <-> board.from );
   not board.to.type = p.type;
   distance(from,to) ={ M,N} ;
   ( abs (M)=1 & abs (N)=2 ) or ( abs (M)=2 & abs (N)=1 );
   begin if ( location ( x <-> board.from ) & x.type \= p.type ) then
      retract ( location ( x <-> board.from ) );
   end if
   retract ( location ( p <-> board.from ) );
   assert ( location ( p <-> board.to ) );
end action

begin action queen ( ptype p, square from, square to )
   p.type = queen;
   location ( p <-> board.from );
   not board.to.type = p.type;
   d : = distance ( from, to );
   d ={ _,0} or d ={ 0,_} or ( d ={ M,N} & abs (M) = abs (N) );
   begin foreach square X in between (board,from,to) :
      not location ( X );
   end foreach
   begin if ( location ( x <-> board.from ) & x.type \= p.type ) then
      retract ( location ( x <-> board.from ) );
   end if
   retract ( location ( p <-> board.from ) );
   assert ( location ( p <-> board.to ) );
end action

begin action bishop ( ptype p, square from, square to )
   p.type = bishop;
```

```
    location ( p <-> board.from );
    not board.to.type = p.type;
    distance (from, to) ={ M,N} ; abs (M) = abs (N);
    begin foreach square X in between (board, from, to) :
       not location ( X );
    end foreach
    begin if ( location ( x <-> board.from ) & x.type \= p.type ) then
       retract ( location ( x <-> board.from ) );
    end if
    retract ( location ( p <-> board.from ) );
    assert ( location ( p <-> board.to ) );
end action

begin action castle ( ptype p, square from, square to )
    location ( p <-> board.from );
    not board.to.type = p.type;
    d : = distance (from, to); d ={ _,0} or d ={ 0,_} ;
    begin foreach square X in between (board, from, to) :
       not location ( X );
    end foreach
    begin if ( location ( x <-> board.from ) & x.type \= p.type ) then
       retract ( location ( x <-> board.from ) );
    end if
    retract ( location ( p <-> board.from ) );
    assert ( location ( p <-> board.to ) );
end action
```

Appendix B

Kassala [2, chap 4].

```
type player = christian | moslem;

begin map kassala
type hexagonal 8 8;
coordinates "0Nou.+90.0N";
terrain loc clear;
terrain loc kassala;
terrain loc udaka;
terrain dir wadi;
terrain dir trench;

clear: 0104, 0105, 0106, 0107, 0108, 0201, 0202, 0203, 0204, 0205,
0206,  0207, 0208, 0302, 0303, 0304, 0306, 0307, 0308, 0401, 0402,
0403,  0404, 0405, 0406, 0407, 0408, 0501, 0502, 0503, 0504, 0505,
0506,  0507, 0508, 0601, 0602, 0603, 0604, 0605, 0606, 0607, 0608,
0702,  0703, 0704, 0705, 0706, 0707, 0708, 0802, 0803, 0806, 0807,
0808;
kassala: 0804, 0805;
```

```
udaka: 0305;
wadi: 0208 >0308, 0207 >0208, 0207 >0307, 0206 >0307, 0206 >0306,
0205 >0306, 0205 >0305, 0204 >0305, 0204 >0304, 0203 >0304, 0203
>0303, 0202 >0303, 0302 >0303, 0302 >0402, 0401 >0402, 0302 >0402,
0406 >0407, 0507 >0407, 0508 >0407;
trench: 0404 >0304, 0404 >0403, 0504 >0403, 0504 >0503, 0603 >0503,
0603 >0602, 0603 >0703, 0704 >0703, 0704 >0803;
end map

define unit{ player : player, type : type, str : int, mf : int, mfmax :
int, haf : bool} ;
type type = infantry | cavalry | cannon;
unit GE1{ player=christian, type=infantry, str=4, mfmax=2} ;
unit GE2{ player=christian, type=infantry, str=4, mfmax=2} ;
unit E1{ player=christian, type=infantry, str=4, mfmax=2} ;
unit E2{ player=christian, type=infantry, str=4, mfmax=2} ;
unit P1{ player=christian, type=infantry, str=4, mfmax=2} ;
unit P2{ player=christian, type=infantry, str=4, mfmax=2} ;
unit PS{ player=christian, type=infantry, str=2, mfmax=2} ;
unit CV{ player=christian, type=cavalry, str=2, mfmax=3} ;
unit CC{ player=christian, type=cannon, str=1, mfmax=0} ;
unit M1{ player=moslem, type=cavalry, str=3, mfmax=3} ;
unit M2{ player=moslem, type=cavalry, str=3, mfmax=3} ;
unit A1{ player=moslem, type=cavalry, str=3, mfmax=3} ;
unit A2{ player=moslem, type=cavalry, str=3, mfmax=3} ;
unit TC{ player=moslem, type=cavalry, str=4, mfmax=3} ;
unit TI{ player=moslem, type=infantry, str=4, mfmax=2} ;
unit N1{ player=moslem, type=infantry, str=3, mfmax=2} ;
unit N2{ player=moslem, type=infantry, str=3, mfmax=2} ;
unit N3{ player=moslem, type=infantry, str=3, mfmax=2} ;
unit ES{ player=moslem, type=infantry, str=4, mfmax=2} ;
unit MC1{ player=moslem, type=cannon, str=1, mfmax=0} ;
unit MC2{ player=moslem, type=cannon, str=1, mfmax=0} ;
unit MC3{ player=moslem, type=cannon, str =1, mfmax=0} ;

relation location : unit <-> hex;
location : GE1 <-> kassala.0804; location : GE2 <-> kassala.0805;
location : E1 <-> kassala.0803; location : E2 <-> kassala.0602;
location : P1 <-> kassala.0304; location : P2 <-> kassala.0305;
location : PS <-> kassala.0403; location : CV <-> kassala.0502;
location : M1 <-> kassala.0608; location : M2 <-> kassala.0708;
location : A1 <-> kassala.0607; location : A2 <-> kassala.0707;
location : TC <-> kassala.0408; location : TI <-> kassala.0308;
location : N1 <-> kassala.0107; location : N2 <-> kassala.0206;
location : N3 <-> kassala.0207; location : ES <-> kassala.0407;
location : CC <-> kassala.0503; location : MC1 <-> kassala.0307;
location : MC2 <-> kassala.0307; location : MC3 <-> kassala.0307;

type cmbt_res =C | EX | AE | DE;
```

```
begin function movement_cost ( unit u, hex from, hex to ) = int
   variable s : side ;
   s := terrain(from,to);
   begin if ( terrain(s) = wadi | terrain(s) = trench ) then
      return(2);
   else
      return(1);
   end if
end function

begin action move ( player p, unit u, hex to )
   u.player = p;
   not contact (u);
   variable from : hex;
   variable cost : int;
   is_instance( location (u <-> from ) );
   distance(kassala,from,to) = 1;
   not impassable(u,from,to);
   cost = movement_cost(u,from,to);
   not cost > u.mf;
   do_move(u,from,to);
   begin if contact(u) then
      u.mf := 0;
   else
      u.mf =- cost;
   end if
end action

begin function impassable ( unit u, hex from, hex to ) = bool
   return ( u.type= cavalry
      & ( terrain ( to )=kassala | terrain(to)=udaka | terrain
      (from,to)=wadi ) );
end function

begin procedure do_move ( unit u, hex from, hex to )
   retract ( location ( u <-> from ) );
   assert ( location ( u <-> to ) );
end procedure
begin function contact ( unit u ) = bool
   variable b : bool := false;
   variable e : player := enemy(u.player );
   variable x : hex;
   is_instance ( location( u <-> x ) );
   begin foreach hex n in neighbours(x) :
      begin if ( is_instance(location(u<->n)) & u.player=e ) then
         b := true;
   end if
end foreach
   return(b);
end function
```

```
begin action combat ( player p, set(unit) as, set(unit) ts )
   begin foreach unit a in as : a.haf = false; end foreach
   begin foreach unit t in t s : u.haf = false; end foreach
   begin foreach unit aa in as :
      begin foreach unit tt in t s :
         variable al, tl : hex;
         is_instance( location( aa <-> al ) );
         is_instance( location( tt <-> tl ) );
         distance (kassala,al,tl) = 1;
      end foreach
   end foreach
   variable hs : set(hex) := empty;
   begin foreach unit tx in ts :
      variable x : hex;
      is_instance ( location( x <-> t x ) );
      insert(x,hs);
   end foreach

begin case crt[ basic_ratio(as,ts)][ random(1,6)]
   DE : eliminate(ts); advance(as,hs);
   AE : eliminate(as);
   EX : exchange(as,as,ts);
      begin if not empty(as) then
        advance(as,hs);
      end if
   C : begin end
   end case
   begin foreach unit u in as : u.haf:=true; end foreach
   begin foreach unit ttt in ts : ttt.haf:=true; end foreach
end action

begin function attack_modifier ( set (unit) as, set(unit) ts ) = int
   variable m : int := 0;
   begin foreach unit a in as:
      begin foreach unit t in ts :
         begin if directed (location(a), location(t)) then
           m := m + 1;
         end if
      end foreach
   end foreach
    return(m);
end function

begin function defense_modifier ( set (unit) ts ) = int
   variable m : int := 0;
   begin foreach unit t in ts :
      variable x : terrain := terrain (location(t));
      begin if x = kassala then
        m := m + 2;
      else if x = udaka then
```

```
            m := m + 1;
          end if
        end foreach
        return(m);
    end function

    begin function basic_ratio ( set(unit) as, set(unit) ts ) = float
        variable astr, dstr : int;
        begin foreach unit a in as : astr =+ a.str; end foreach
        begin foreach unit t in ts : dstr =+ t.str; end foreach
        astr =+ attack_modifier(as,ts);
        dstr =+ defense_modifier(ts);
        return(ratio(astr,dstr));
    end function

    begin procedure exchange ( set(unit) as, set(unit) adj, set(unit) ts )
        variable astr, dstr : int;
        begin foreach unit a in as : astr=+a.str; end foreach
        begin foreach unit t in ts : dstr=+ t.str; end foreach
        begin if (astr > dstr) then
            variable rest : set(unit) := empty;
            eliminate(ts);
            pick_elim(as,dstr);
            begin foreach unit aa in as :
              begin if ( is_instance( location( aa <-> _ ) ) ) then
                  insert(aa,rest);
              end if
            end foreach
            advance(rest,??);
        else
            eliminate(as);
            pick_elim(ts,astr);
        end if
    end procedure

    begin procedure eliminate ( set(unit) us)
        begin foreach unit u in us :
            retract( location ( u <-> _ ) );
        end foreach
    end procedure

    begin procedure pick_elim ( set (unit) us, int str )
        variable u : unit;
        begin loop
            dialogue select us u;
            retract ( location( u <-> _ ) );
            delete(u,us);
            str := (str – u.str);
            begin if str < 0 | empty(us) then
              exit;
```

```
        end if
     end loop
  end procedure

  begin procedure advance ( set(unit) us, set(hex) hs )
     begin foreach hex x in hs :
        dialogue select ??
     end foreach
  end procedure

  variable result : winner;
  type winner = christian | moslem | draw;

  begin function christians_in( terrain t ) = set(unit)
     variable us : set(unit);
     begin findall unit u = us :
        variable x : hex;
        is_instance( location ( u <-> x ) );
        terrain(x) = t;
        u.player = christian;
     end findall
        return(us);
  end function

  begin function stacking = bool
     variable b : bool : = true;
     variable us : set(unit);
     begin foreach hex x :
        begin find all unit u = us :
          is_instance ( location ( u <-> x ) );
          not u.type = cannon;
        end findall
        begin if (count(us) > 1) then b := false; end if
     end foreach
     return(b);
  end function

  begin function mustfight = bool
     variable b : bool := true;
     begin foreach unit u :
        begin if (contact(u) & u.haf = false) then
          b := false;
        end if
     end foreach
        return(b);
  end function

  begin sop
     variable winner : winner;
     variable ck, cu : set(unit);
```

```
    begin for[ i:int, 1.. 10]
       begin foreach unit u : u.mf:=u.mfmax; u.haf:=false; end
       foreach
       dialogue many{ move(moslem)} { stacking} ;
       dialogue many{ combat(moslem)} { mustfight} ;
       dialogue many{ move(christian)} { stacking} ;
       dialogue many{ combat(christian)} { mustfight} ;
    end for
    ck = Christians_in(kassala);
    cu = Christians_in(udaka);
    begin if ( empty(ck) & empty(cu) ) then
       winner := moslem;
    else if ( not empty(ck) & not empty(cu) ) then
       winner := christian;
    else
       winner := draw;
    end if
end sop

begin table crt[ ratio[ 1/4|5/1] x int[ 1-6] => cmbt_res]
"1-4<" "1-3" "1-2" "1-1" "2-1" "3-1" "4-1" "5-1>"
1   C    EX   EX   C    DE   DE   DE   DE
2   C    C    EX   EX   C    DE   DE   DE
3   AE   C    C    EX   EX   C    DE   DE
4   AE   AE   C    C    EX   EX   C    DE
5   AE   AE   AE   C    C    EX   EX   C
6   AE   AE   AE   AE   AE   C    EX   EX
end table
```

Appendix C

Drive on Metz [3, pp. 174–89].

```
type player = German | "U.S.";
variable result : winner;

type winner = German | "U.S." | draw;
variable metz : player := German;
variable thionville : player := German;
variable eotm0, eotm1, eotm2, eotm3 : set (unit) := empty;
variable us_vp, ger_vp : int := 0;

begin map metz
type hexagonal 9 11;
coordinates "0N.-90.0Noi";

terrain loc clear;
terrain loc forest;
terrain loc rough;
```

```
terrain extra town;
terrain extra fortified;
terrain c2c road;
terrain side river;

clear: 0101, 0102, 0103, 0104, 0106, 0107, 0108, 0109, 0110, 0111,
0201, 0203, 0204, 0205, 0207, 0210, 0211, 0304, 0405, 0406, 0407,
0506, 0511, 0601, 0607, 0701, 0702, 0703, 0704, 0705, 0706, 0707,
0708, 0709, 0710, 0801, 0802, 0803, 0804, 0805, 0806, 0807, 0808,
0809, 0901, 0902, 0905, 0906, 0907, 0908;
forest: 0202, 0301, 0302, 0303, 0305, 0310, 0311, 0401, 0402, 0403,
0404, 0408, 0409, 0410, 0411, 0502, 0503, 0504, 0505, 0508, 0509,
0510, 0602, 0603, 0604, 0605, 0606, 0608, 0903, 0904, 0909, 0910;
rough: 0306, 0307, 0308, 0309, 0501, 0609, 0610, 0611, 0711, 0810,
0811, 0911;
river: 0701|0801, 0702|0801, 0702|0802, 0703|0802, 0703|0803,
0704|0803, 0804|0803, 0804|0904, 0804|0905, 0805|0905, 0805|
0906, 0806|0906, 0806|0907;
river: 0806|0807, 0707|0807, 0707|0708, 0607|0708, 0608|0708,
0608|0709, 0608|0609, 0509|0609, 0510|0609, 0510|0610, 0511|
0610, 0511|0611;
river: 0807|0907, 0807|0908, 0807|0808, 0708|0808, 0709|0808,
0809|0808, 0809|0909, 0809|0910, 0809|0810, 0710|0810, 0711|
0810, 0711|0811;
town: 0105 (Abbeville), 0206 (Jarny), 0208 (Mars la Tours), 0209
(Chambley), 0302 (Trieux), 0304 (Briey), 0311 (Rupt de Mad), 0409
(Gorze), 0506 (St Privat), 0507 (Gravelotte), 0509 (Dornot),
0510 (Arnaville), 0701 (Thionville), 0807 (Metz); fortified:
0507, 0508, 0509, 0606, 0609, 0702, 0709, 0806;
road: 0103–0203, 0104–0203, 0203–0304, 0204–0304, 0205–0204,
0206–0205, 0106–0206, 0207–0206, 0208–0207, 0108–0208,
0209–0208, 0110–0209, 0210–0209, 0310–0210, 0311–0310,
0211–0311, 0304–0303, 0303–0402, 0402–0502, 0401–0502,
0502–0601, 0601–0701, 0601–0702, 0701–0702, 0702–0703,
0703–0704, 0704–0804, 0304–0305, 0305–0405, 0405–0506,
0506–0605, 0605–0604, 0604–0705, 0705–0804, 0206–0307,
0307–0407, 0407–0507, 0208–0308, 0308–0407, 0506–0507,
0311–0410, 0410–0510, 0511–0510, 0209–0309, 0309–0409,
0409–0510, 0510–0509, 0509–0608, 0608–0508, 0508–0507,
0608–0607, 0507–0607, 0607–0708, 0708–0707, 0707–0807,
0807–0806, 0806–0805, 0805–0804, 0611–0610, 0610–0609,
0609–0709, 0709–0708, 0708–0807, 0811–0810, 0810–0910,
0910–0909, 0909–0808, 0808–0807, 0908–0807, 0807–0907,
0907–0906, 0906–0905, 0905–0904, 0904–0803, 0803–0802,
0802–0801;
end map

relation location : unit <–> ( hex | pool );

object pool;
```

```
define unit{ player:player, str:int, mp:int, mpmax:int, haf:
bool};
unit "559/1125"{ player=German, str=1, mpmax=4};
unit "559/1126"{ player=German, str=1, mpmax=4};
unit "462/UTRFHR"{ player=German, str=2, mpmax=4};
unit "462/1010"{ player=German, str=1, mpmax=4};
unit "462/FHNJKR"{ player=German, str=3, mpmax=4};
unit "3PG/8PG"{ player=German, str=2, mpmax=8};
unit "3PG/29PG"{ player=German, str=2, mpmax=8};
unit "17SS/3855"{ player=German, str=2, mpmax=8};
unit "17SS/3755"{ player=German, str=3, mpmax=8};
unit "Metz gar."{ player=German, str=1, mpmax=1};
unit "106PzB"{ player=German, str=1, mpmax=8};
unit "90/358"{ player="U.S.", str=4, mpmax=4};
unit "90/357"{ player="U.S.", str=4, mpmax=4};
unit "7A/CCA"{ player="U.S.", str=7, mpmax=10};
unit "5/2"{ player="U.S.", str=5, mpmax=4};
unit "7A/CCR"{ player="U.S.", str=5, mpmax=10};
unit "7A/CCB"{ player="U.S.", str=7, mpmax=10};
unit "5/11"{ player="U.S.", str=5, mpmax=4};
unit "5/10"{ player="U.S.", str=5, mpmax=4};

begin table reinforcement[ unit |> set(hex)]
"90/358"{ 0101, 0102, 0103, 0104}
"90/357"{ 0101, 0102, 0103, 0104}
"7A/CCA"{ 0105, 0106, 0107}
"5/2"{ 0105, 0106, 0107}
"7A/CCR"{ 0108, 0109, 0110, 0111}
"7A/CCB"{ 0108, 0109, 0110, 0111}
"5/11"{ 0108, 0109, 0110, 0111}
"5/10"{ 0108, 0109, 0110, 0111}
"106PzB"{ 0401, 0501, 0601, 0701, 0801, 0901}
end table

location : "559/1125" <-> metz.0502;
location : "559/1126" <-> metz.0403;
location : "462/UTRFHR" <-> metz.0505;
location : "462/1010" <-> metz.0507;
location : "462/FHNJKR" <-> metz.0509;
location : "3PG/8PG" <-> metz.0609;
location : "3PG/29PG" <-> metz.0611;
location : "17SS /3855" <-> metz.0709;
location : "17SS /3755" <-> metz.0808;
location : "Metz gar." <-> metz.0807;

begin sop
   begin for[ i:int, 1..7]
      begin foreach unit u : u.mp:=u.mpmax; u.haf:=false; end
      foreach
      begin if i=1 then
```

```
        assert ( location ( ”90/358” <–> pool ) );
        assert ( location ( ”90/357” <–> pool ) );
        assert ( location ( ”7A/CCA” <–> pool ) );
        assert ( location ( ”5/2” <–> pool ) );
        assert ( location ( ”7A/CCR” <–> pool ) );
        assert ( location ( ”7A/CCB” <–> pool ) );
        assert ( location ( ”5/11” <–> pool ) );
        assert ( location ( ”5/10” <–> pool ) );
      else if i=2 then
        assert ( location ( ”106PzB” <–> pool ) );
      end if
      dialogue many{ move (”U.S.”),enter (”U.S.”),exit(i,
      ”U.S.”)}{ stacking} ;
      dialogue many{ combat (”U.S.”)} ;
      dialogue many{ move (German),enter(German),exit(i,
      German)} { stacking} ;
      dialogue many{ combat ( German ) } ;
      begin findall unit u = eotm :
        u.player = ”U.S.”;
        location ( u <–> x );
        member(x,[ 0801,0901,0802,0902,0803,0903,0904,0905,
                0906,0807,0907, 0708,0808,0908,0609,0709,
                0809,0909,0610,0710,0810,0910,0611, 0711,
                0811,0911] );
      end findall
      eotm3 := intersection(union(eotm2,eotm3),eotm);
      eotm2 := intersection(union(eotm1,eotm2),eotm);
      eotm1 := intersection(union(eotm0,eotm1),eotm );
      eotm0 := difference(eotm,union(eotm1,eotm2,
      eotm3 ) );
    end for
    result := winner(ger_vp,us_vp);
  end sop

  begin action move ( player p, unit u, hex to )
    variable from : hex;
    variable cost : int;
    u.player = p;
    u.mp > 0;
    location ( u <–> from );
    distance(metz,from,to) = 1;
    not zoc(u,from);
    cost := movement_cost(u,from,to);
    not cost > u.mp;
    do_move(u,to);
    begin if zoc(u,to) then
      u.mp := 0;
    else
      u.mp =- cost;
    end if
```

```
      record_move(p,to);
   end action

   begin procedure do_move ( unit u, hex to )
      variable from : hex;
      location ( u <—> from );
      retract ( location ( u <—> from ) );
      assert ( location ( u <—> to ) );
   end procedure

   begin function movement_cost ( unit u, hex from, hex to ) = int
      begin if c2c (from,to,road) then
         return (1);
      else if hexside (from,to,river) then
         return (u.mp);
      else
         return(tec_m[ terrain(to)] );
      end if
   end function

   begin function stacking = bool
      variable b : bool := true;
      variable us : set (unit);
      begin foreach hex x :
         begin findall unit u = us :
            location ( u <—> x );
         end findall
         begin if (count(us) > 1) then b := false; end if
      end foreach
      return (b);
   end function

   begin action combat ( player p, set(unit) us, unit t )
      location ( t <—> x );
      subset( us, neighbours(t) );
      begin foreach unit u in us : u.player = p; u.haf=false; end
      foreach
      t.haf=false;
      variable astr, diff : int;
      begin foreach unit a in us : astr =+ a.str; end foreach
      diff = astr—t.str;
      ??? tec_c[ terrain] ;
      begin case crt[ diff][ random(1 ,6)]
         AR : begin foreach unit vv in us : retreat (vv);
         end foreach
         DR : retreat(t); advance (us,x);
         DR2 : begin for each unit vv in us :
                     retreat2 (vv); advance (us,x);
               end foreach
      end case
      begin foreach unit v in us : v. haf:= true; end foreach
```

```
      t.haf:=true;
   end action

   begin function winner = winner
      us_vp =+ length(eotm3) * 5;
      begin if (0701.owner="U.S.") then us_vp =+ 5; end if
      begin if (0807.owner="U.S.") then us_vp =+ 20; end if
      variable int diff := ger_vp — us_vp;
      variable winner w;
      begin if (diff > 15) then w := German decisive
      else if (diff > 10) then w := German substantial
      else if (diff > 5) then w := German marginal
      else if (diff < —5) then w := US marginal
      else if (diff < —10) then w := US substantial
      else if (diff < —15) then w := US decisive
      else w := draw;
      return (w);
   end function

   begin function us_exit_vp ( unit u, hex x ) = int
      begin if member (x,[ 0901,0902,0903,0904,0905,0906,0907,
      0908,0909,0910,0911] ) then return (5);
      else
         return (0);
      end if
   end function

   begin function ger_exit_vp ( unit u, hex x, turn n ) = int
      variable int vp := 0;
      begin if member (x,[ 0101,0102,0103,0104,0105,0106,0107,
      0108,0109,0110,0111] ) then
         vp := 10
      else if ( ( u.division = 3PG | u.division = 17SS )
         & (member (x,[ 0901,0902,0903,0904,0905,0906,0907,0908,
         0909,0910,0911] ) | member (x,[ 0111 ,0211 ,0311 ,041 1 ,0511 ,
         0611 ,0711 ,0811 ,0911] ) )
         ) then
        vp := (8 — turn)
      end if
      return (vp);
   end function

   begin function zoc ( unit u, hex x ) = bool
      variable b : bool := false;
      variable e : player := enemy (u. player);
      begin foreach hex n in neighbours ( x) :
         begin if ( location ( u <—> n ) & u.player=e ) then
            b := true;
         end if
      end foreach
      return (b);
```

```
end function

begin action enter ( unit u, hex x )
    reinforcement[ u] = hexes;
    member(x,hexes);
    not impassable(x,u);
    retract ( location ( u <—> pool ) );
    assert ( location ( u <—> x ) );
end action

begin action exit ( turn n, unit u, hex from )
    location ( u <—> from );
    retract ( location ( u <—> from ) );
    begin case u. player
        ”U.S.” : us_vp =+ us_exit_vp (u,x);
        German : ger_vp =+ ger_exit_vp (u,x,n);
    end case
end action

begin procedure record_move ( player p, hex x )
    begin if x = 0701 then
        thionville = p;
    else if x = 0807 then
        thionville = p;
    end if
end procedure

begin table tec_m[ primary —> int]
clear forest rough
2     4      3
end table

begin table tec_c[ terrain —> int]
clear forest rough town fortified road river
0     2      1     2    3         0    3
end table

type cmbt_res = ”-” | AR | DR | DR2;

begin table crt[  int x int[ 1−6] => cmbt_res ]
       ”−1+”     0      ”+1”    ”+2,+3”   ”+4,+5”   ”+6,+7”   ”+8,+9”   ”+10+”
1      ”−”       DR     DR      DR        DR2       DR2       DR2       DR2
2      ”−” ”−”   DR     DR      DR        DR2       DR2       DR2
3      AR        ”−”    ”−”     DR        DR        DR        DR2       DR2
4      AR        AR     AR      ”−”       DR        DR        DR        DR2
5      AR        AR     AR      R         ”−”       DR        DR        DR
6      AR        AR     AR      AR        AR        ”−”       DR        DR
end table
```

The G-Code Game-Play Formalism
DOI: http://dx.doi.org/10.5772/intechopen.1007672

Author details

Tomas By
Florida Institute for Human & Machine Cognition, Pensacola, FL, USA

*Address all correspondence to: tby@ihmc.us

References

[1] By T. Formalizing game-play, Simulation & Gaming. 2012;**43**(2):157-187

[2] Freeman J. The Complete Book of Wargames. New York, New York: Simon and Schuster; 1980

[3] Dunnigan JF. Wargames Handbook. 3rd ed. Lincoln, Nebraska: Writers Club Press; 2000

[4] Peterson J. Playing at the World. San Diego, California: Unreason Press; 2012

[5] Harrigan P, Kirschenbaum MG, editors. Zones of Control. Cambridge, Massachusetts: MIT Press; 2016

[6] Murray H. A History of Board-Games other than Chess. Oxford: Oxford University Press; 1951

[7] Parlett D. The Oxford History of Board Games. Oxford: Oxford University Press; 1999

[8] By T. Hexagonal grid numbering. IEEE Transactions on Games. 2021;**13** (2):138-145

[9] Card OS. Gameplay, Compute!1989. p. 12

[10] Shapiro NZ, Hall HE, Anderson RH, LaCasse M. The RAND-ABEL programming language: History, rationale, and design, Report R-3274-NA. Santa Monica, California: The RAND Corporation; 1985

[11] Shapiro NZ, Hall HE, Anderson RH, LaCasse M, Gillogy MS, Weissler R. The RAND-ABEL programming language: Reference manual, Note N-2367-1-NA. Santa Monica, California: The RAND Corporation; 1985

[12] Davis PK. An Analyst's Primer for the RAND/ABEL Programming Language, Note N-3042-NA. Santa Monica, California: The RAND Corporation; 1990

[13] Pell B. METAGAME in symmetric, Chess-like games. In: van den Herik HJ, Allis LV, editors. The Third Computer Olympiad, Vol. 3, Heuristic Programming in Artificial Intelligence. Chichester: Ellis Horwood; 1992. pp. 252–280

[14] Pell B. METAGAME: A new challenge for games and learning. In: van den Herik HJ, Allis LV, editors. The Third Computer Olympiad, Vol. 3 Heuristic Programming in Artificial Intelligence. Chichester: Ellis Horwood; 1992. pp. 237–251

[15] Pell B. Logic programming for general game-playing. In: Proceedings of the Workshop on Knowledge Compilation and Speedup Learning at Machine Learning Conference. Amherst, Massachusetts. 1993

[16] Romein J, Bal H, Grune D. Multigame—A very high level language for describing board games. In: First Annual ASCI Conference. 1995. pp. 278-287

[17] Romein JW, Bal HE, Grune D. The Multigame reference manual, Informatica rapport 475. Amsterdam: Vrije Univeristeit; 2000

[18] Kaiser DM. The structure of games [Ph.D. thesis]. Miami, Florida: Florida International University; 2007

[19] Quenault M, Cazenave T. Extended general gaming model. In: Proceedings of the Computer Games Workshop.

Amsterdam: Maastrich University; 2007. pp. 195-204

[20] Maggiore G, Spanò A, Orsini R, Costantini G, Bugliesi M, Abbadi M. Designing Casanova: A language for games. In: Advances in Computer Games. Vol. 7168 of Lecture Notes in Computer Science. Berlin/Heidelberg: Springer; 2012. pp. 320-332

[21] Maggiore G, Spanò A, Orsini R, Bugliesi M, Abbadi M, Steffinlongo E. A formal specification for Casanova, a language for computer games. In: Proceedings of EICS. 2012. pp. 287-292

[22] Silver D, Hubert T, Schrittwieser J, Antonoglou I, Lai M, Guez A, Lanctot M, Sifre L, Kumaran D, Graepel T, Lillicrap T, Simonyan K, Hassabis D. Mastering Chess and Shogi by self-play with a general reinforcement learning algorithm. 2017

[23] Harmon RD. The fallacy of panzerbush. The General. 1975;**12**(2): 25-26

Chapter 2

Lanchester's Heterogeneous "Fighting Strength" as a Battle Outcome Predictor in a Ternary Mixed Force Competition Applied to a Simple Fire and Manoeuvre Wargame

Mark Flanagan, David Lambert, Trevor C. Lipscombe, Adrian Northey and Ian M. Robinson

Abstract

This study extends our previous work on battle-outcome prediction using a simple fire-and-manoeuvre wargame, leading to force annihilation. Ternary mixtures explore how properties change by mixing three components in various ratios to identify ideal compositions. We use this approach in a Double Round-Robin competition composed of mixtures of three types of forces (good, average and poor) which financially constrained the number and type of forces used. The wargame outcomes are explored using Lanchester's heterogeneous equations, and by a salvo based model, which is governed to the cumulative binomial distribution for rolling 1D6 dice repeatedly. An optimal force balance is identified within the study.

Keywords: Lanchester's equations, heterogenous fighting strength, salvo, cumulative binomial distribution, experiment with mixtures, wargame

1. Introduction

The sword devoureth one as well as another

2 Sam 11:25

Despite the stochastic element of battles, the outcomes can have an underlying predictability. In our earlier studies [1–5] we demonstrated that Lanchester's equations can be used to predict the outcome of battles fought between a Blue and Red force in a simple manoeuvre and fire wargame—Möbius Mayhem. The paper derived Lanchester's 'Fighting Strength' in the square law relationship

$$\text{`Fighting Strength'} = P\,N^2 \quad (1)$$

where P is the 'efficiency' or probability of hitting an opponent in a fire exchange, removing them from further combat, and N is the number of troops available to fight in the fire exchange. Lanchester's theory assumes simultaneous and continuous fire exchange in the battle leading to a steady reduction in numbers of combatants for each side. In the 'Möbius Mayhem' wargame (MMW) used in the study players use a fixed turn sequence (IgoYougo), leading to sequential salvo firing, making the game akin to a Markov Chain. We predicted the final result in 33 out of 34 wargames with asymmetric forces (with either different number of troops & quality), and detailed further games where the forces were symmetrical in terms of number of troops & quality. Using solutions to balance Fighting Strengths between forces with different fighting efficiencies or numbers, nearly symmetric battles were also fought. Stochastic behaviour in fire exchange leads to the balance of forces swinging to one side, and thereafter, one side is victorious. A slight bias was detected between the symmetric and nearly symmetric forces with 60% of battles fought won by sides with 1st salvo advantage. In addition Lanchester's equations also gave good estimates for the % number of casualties and the length of time in each wargame.

The game is described fully in Refs. [2–5], but briefly players used a playing surface which was a representation of a Möbius strip. There was a fixed turn sequence and the sequence of play was Move, then Fire. They could move up to 5 hexes per turn, and could pass through the same hex as another unit, but may not remain in the same hex as another unit at the end of the turn. Trapdoors were available for rapid movement consistent with the surface. Units that moved could not fire and likewise units that fired could not move, and a clear line of sight was needed to fire. To hit and eliminate an enemy with regular quality troops, count the distance in hexes, and a 1D6 was rolled, with the score having to match or exceed the range in hexes, with 6 hexes maximum range. Troop quality also varied. Good (G) troop quality had a (+1) modifier, Average (A) troop quality had (0) modifier and Poor (P) troop quality had a (−1) modifier applied to all 1D6 throws. This affected the probability of hitting and eliminating an enemy as a function of range (**Table 1**).

Consequently, at a range of 6 hexes, a roll of 6 is needed for average quality troops, or 5, 6 for good quality troops. Poor quality troops cannot hit at a range of 6 hexes, with effective combat from 5 hexes or less. Battles were normally conducted at a range where mutual fire exchange could occur (for combinations of good and average = 6 hexes, or for those involving poor quality troops = 5 hexes).

In this paper we explore MMW with heterogeneous force compositions in a Double Round-Robin Format, with allocation of Good, Average and Poor troop quality of differing numbers, resulting in heterogeneous force compositions within a cost constraint. Each troop type fights all others, in one round with 1st salvo advantage and in the second round

1D6 modifier	1 hex	2 hex	3 hex	4 hex	5 hex	6 hex
(+1)	1	1	5/6	2/3	1/2	1/3
0	1	5/6	2/3	1/2	1/3	1/6
(−1)	5/6	2/3	1/2	1/3	1/6	0

Table 1.
Probability, P, vs. range in hexes to secure a hit and remove of an enemy unit for troops of good quality (+1), regular quality (0) and poor quality (−1).

with the salvo advantage reversed. Thus the competition is fair (all forces fight all others) and balanced (equal number of 1st and 2nd salvo advantage fights). The battles in this paper are explored using Lanchester's equations solved for heterogeneous forces, leading to the concept of a heterogeneous fighting strength (HFS). It is also explored using a salvo based model, leading to the concept of 'Salvo Intensity'.

2. Theoretical considerations of heterogeneous force combat

2.1 Lanchestrian (n,m) models

Understanding combat with heterogeneous forces is a classic problem studied using Lanchester's equations. Models have been presented for studying heterogeneity on both sides, with m types of Red unit facing n types of blue unit (the (n, m) model). Each n,m have differing numbers and probability of hitting their opponent. The first, due to Lanchester [6] states: *'Where the component units differ among themselves, as in the case of a fleet that is not homogeneous, the measure of the total of fighting strength of a force will be the square of the sum of the square roots of the strengths of its individual units.'* so that

$$\text{Fighting Strength} = \left(\sum\nolimits_{i=1}^{n} \sqrt{P_i N_i}\right)^2 \tag{2}$$

where P and N are the probability of hitting and N the number of the ith component in the mixed force. This was offered without proof as pointed out by Lepingwell [7], who derived another form for the Heterogeneous Fighting Strength (HFS).

Colegrave and Hyde solved the Lanchester (2, 2) problem using Hamiltonian equations [8]. Mackay [9] studied mixed forces under aimed fire with mixed forces on both one side and mixed sides, and for best target allocation under these circumstances. The equations for Blue mixed (i types of forces) vs. Red homogenous are

$$\frac{dB_i}{dt} = -r^i \mu^i R, \frac{dR}{dt} = -\sum\nolimits_{i=1}^{n} b_i B_i \tag{3}$$

where the terms r, and, b are kill-rate probabilities, B, R are the number and type (for blue) of troops and μ is a term introduced for target assignment. Solutions to this can lead to Lepingwell's formulation on certain assumptions. Mackay extended the analysis to mixed forces and unaimed fire. Regarding target allocation the following is noted:

Red's optimal strategy is for type α to rank the opposing units by $r_\alpha^1 r_1^\alpha < r_\alpha^2 b_2^\alpha < \ldots r_\alpha^n b_n^\alpha$ *and to fire on each type in succession, from n down, until it sees it annihilated (and similarly for Blue).* Thus it is essential to target the most powerful troop component first in a mixed force to ensure optimal survival. This aspect was further explored by Lin and MacKay [10].

2.2 Salvo based models

Armstrong [11] reviewed Salvo Models for Naval Surface Combat, which includes an intercepting component of one force against any attacks (missiles in terms of modern naval warfare).

$$\Delta A = -(\beta B - yA)/w, \Delta B = -(\alpha A - zB)/x \tag{4}$$

with the changes in forces A, B by salvo dependent upon the number of attacks per salvo (α, β) and counter attacks to this (y,z), with the number of hits to damage a ship beyond combat potential or sink it being (x,w). Armstrong [12] extended this analysis to include stochastic elements.

2.3 Stochastic models

Taylor [13] looked at system states and transition probabilities in a Lanchestrian fire exchange and all the possible outcomes as the states of the Markov Chain model. Assuming that there are three possible outcomes after transition: Blue decreases, Red decreases, or no change, it is possible analytically to introduce stochastics into the analysis.

Mangulis [14] looked at probabilistic salvo fire between forces with their equations for force exchange being:

$$\frac{dy}{dt} = -1\left(1 - \left(\frac{P_x}{y}\right)^x\right)\frac{y}{T_x}, \frac{dy}{dt} = -1\left(1 - \left(\frac{P_y}{x}\right)^y\right)\frac{x}{T_y} \tag{5}$$

where x, y are the numbers of troops for the opposing forces, $P_{x,y}$ are the probabilities of hitting an opponent in fire exchange and $T_{x,y}$ are the step time intervals for fire exchange and loss. These equations in the limit of small probabilities of hitting for $P_{x,y}$ reduce to Lanchester's equations.

$$\left(\frac{x}{y}\right)^2 = \frac{(P_y T_x)}{(P_x T_y)} \tag{6}$$

Kim et al. [15] extended the earlier model by Taylor [13] in introducing stochastic solutions to Lanchester's equations. Kim reformulated this model as stochastic difference equations with Bernoulli random variables, the probability of which depends on the attrition rate and the strength of the opponents. In turn they produced computationally approximate solutions to these variables.

2.4 Comments on theory and MMW

We noted in our earlier paper [2] in Table 3 the differences between Lanchester based models and MMW. The fire exchange in MMW is salvo based, inherently probabilistic, and follows a Markov Chain. Lanchestrian approaches are useful, despite having deterministic and continuous fire exchange. Appendix A shows our derivation of the HFS, which is the same form as Lepingwell's. In terms of MMW this becomes

$$(P_G N_G + P_A N_A + P_P N_P)(N_G + N_A + N_P) \tag{7}$$

Note that if only one type of troop quality is present, Eq. (7) reduces back to the homogenous fighting strength relationship, i.e. 'Fighting Strength' = P N^2. This is used in the paper to calculate HFS and an excel spreadsheet is available for MMW [16] for this purpose. We use Eq. (1) based on Lepingwell to calculate the Heterogeneous

Fighting Strength (HFS) and Mackay's observations on priority target allocation in MMW.

Armstrong's salvo model has limited applicability in MMW, given there is no counter fire to any attack (y = z = 0) and a single hit (x = w = 1) eliminates an individual troop in the game. In Appendix B we present our own Salvo based model which leads to the concept of Salvo Intensity involving average probabilities in fire exchange between heterogenous forces.

In terms of stochastic models, looking at Mangulis is instructive. The probabilities of hitting an opponent in fire exchange $P_{x,y}$ depend upon troop quality and range according to **Table 1**. Success in hitting in turn is the result of 1D6 dice rolls. In Appendix C we present tables of the cumulative binomial distribution probabilities [17] for either a 6, 5,6 or 4-6 on repeated 1D6 rolls up to 20, which matches MMW. One can see from the tables that enhanced success in hitting arises with higher chances (compare the odds for 4-6 vs. 6) and more rolls (= more troops in MMW firing), which underpins Lanchester's concept of fighting strength depending on probability and numbers of troops. However, the outcomes still have a chance of obtaining zero hits, especially as the number of troops begin to reduce.

3. Exploring heterogeneous forces using MMW

3.1 Force variations using a ternary mixtures approach

Our first study used homogenous forces of Good, Average and Poor troops with differing numbers and their combat potential within MMW was explored by the fighting strength, P N^2. 10 Good quality troops are very close to the fighting strength of 14 Average troops at a mutual fire exchange range of 6 hexes ($\frac{1}{3}10^2 \sim \frac{1}{6}14^2 \sim 33$), and similarly 10 Good quality troops are very close to the fighting strength of 17 Poor troops at a range of 5 hexes ($\frac{1}{2}10^2 \sim \frac{1}{6}17^2 \sim 50$). A ternary mixture can be made using 10 Good, 14 Average and 17 Poor quality troops as a test of whether a better force combination can be found than 10 Good quality troops. To constrain the choices of force allocations, an arbitrary budget of $30.00 per force is chosen, with Good troops costing $3.00 each, Average troops $2.14 each and Poor troops costing $1.76 each. Force compositions of any type of number of Good, Average or Poor troops can be made provided the overall cost of the force does not exceed $30.00. We have identified 81 possible combinations of varying troop numbers and qualities that come close to, but do not exceed the budget of $30.00. Experimental design can help reduce this number to better understand the balance of forces using a ternary simplex design [18]. The following forces were chosen to understand this ternary space, and to explore systematically the force composition variations on the HFS and battle outcomes. There are vertices (1,0,0) (0,1,0) (0,0,1), midpoints (½,½,0), (0,½,½), (½,0,½) and the centre point (⅓,⅓,⅓) using the troops identified with 7 forces in the double round-robin tournament. These are shown in **Figure 1** for the number of Good, Average and Poor (G-A-P).

The costs for these heterogeneous forces in this study are shown in **Table 2**.

The forces on the top right side of **Figure 1** are 10-0-0, 5-7-0 & 0-0-14 and none have poor quality troops. The forces on the bottom side are 0-0-17, 0-7-8 & 0-0-14 and none of these forces have any good quality troops. The forces on the top left side are 0-0-17, 6-0-6 & 10-0-0 and none have average quality troops. The force in the middle of the diagram 4-4-5 is a mixture of all three types of troops.

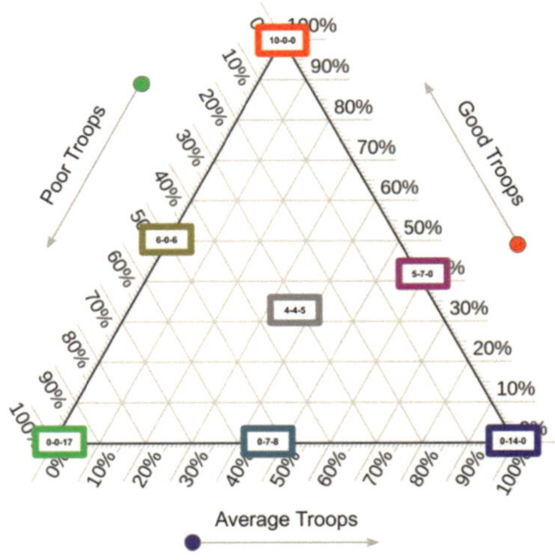

Figure 1.
Ternary simplex mixtures in the study showing the number of Good-Average-Poor troops in the force. The vertex colours are Red (all Good troops), Blue (all Average troops) and Green (all Poor troops) with the colours in between showing the compositional mixtures between the heterogeneous forces based on their fraction of the whole.

Simplex design	Composition (G-A-P)	Good (+1 1D6 modifier)	Average (0 1D6 modifier)	Poor (−1 1D6 modifier)	Cost $
(1,0,0)	10-0-0	10	0	0	30.00
(½,½,0)	5-7-0	5	7	0	29.98
(0,1,0)	0-14-0	0	14	0	29.96
(0,½,½)	0-7-8	0	7	8	29.06
(0,0,1)	0-0-17	0	0	17	29.92
(½,0,½)	6-0-6	6	0	6	29.06
(⅓,⅓,⅓)	4-4-5	4	4	5	29.36

Table 2.
Simplex design parameters, and compositions for homogeneous and heterogeneous forces composed of Good—Average—Poor troop quality and their costs in the ternary mixtures.

Our derivation of Lanchester's HFS in Eq. (7) has been used to determine the precombat HFS [17] for these forces against their paired opponent, and are shown in Appendix D. Note that there is no unique value of HFS, it depends on the number and type of troops and the range of mutual combat, and consequently changes turn by turn in the game as attrition reduces the number of troops.

3.2 Double round-robin competition

3.2.1 Battle outcomes

The mixed forces are arranged with similar troop quality together facing their opponents most lethal troop quality, thus optimising the forces chance of survival.

Battles were fought as a *Vernichtungsschlacht—Battle of Annihilation* with forces moving immediately to optimal mutual firing range and destroying the opposing force. Each unit in the force engaged its opponent in single 1D6 rolls in direct fire. Note that in the games, due to local force annihilation in the combat firing line, movement to re-engage targets was necessary, breaking Lanchester's assumptions of continuous firing. Given the game plays with a fixed turn sequence, an attacker must move into mutual attacking range before firing, necessarily exposing themselves to fire exchange before returning fire in the next turn. In this paper for all battles the attacker advanced on the defender, resulting in the defender with first salvo advantage, with the attacker replying second, and so on turn by turn until one side is annihilated. The outcomes for each battle are shown in Appendix D, **Tables D1–D3** and are available as animated GIFS in Ref. [19]. Each force fights 6 battles with 1st salvo advantage and 6 battles with the 2nd salvo in the double round-robin competition. The tables give the pre-contact full strength HFS, together with the results of the battles fought for all paired forces in both 1st and 2nd salvo advantage fights. All battles are available as animated GIFS, together with plots of the % surviving units, turn by turn. The summary of the battle outcomes are given in **Tables 3** and **4**.

Force (5-7-0) won the competition with the most wins, fewest losses and the best balance of average % survivors after each battle. It had the highest HFS for all forces it

Troop type G-A-P	1st Salvo Wins	1st Salvo Losses	2nd Salvo Wins	2nd Salvo Losses	Total Wins	Total Losses
5-7-0	4	2	6	0	10	2
10-0-0	5	1	4	2	9	3
0-14-0	5	1	2	4	7	5
4-4-5	3	3	4	2	7	5
6-0-6	3	3	2	4	5	7
0-7-8	3	3	0	6	3	9
0-0-17	1	5	0	6	1	11

Table 3.
*Battle outcomes for each troop type in the study indicating wins and losses from the results in **Tables D1** and **D2**.*

Troop type G-A-P	1st Salvo % survivors over all battles	2nd Salvo % survivors over all battles
5-7-0	37.6	51.4
10-0-0	28.3	23.3
0-14-0	60.7	15.4
4-4-5	24.4	35.9
6-0-6	27.8	16.7
0-7-8	25.6	0.0
0-0-17	7.9	0.0

Table 4.
*Average % survivors for each troop type in the study in 1st and 2nd Salvo exchanges from the results in **Tables D1** and **D2**.*

fought. It won both 1st and 2nd salvo advantage battles against the force composed of all Good troops (10-0-0). Given force (5-7-0) had a HFS of 34 and force (10-0-0) had a HFS of 33, this is expected by theory. Force (10-0-0) finished a clear second, losing twice to (5-7-0) and once to (0-14-0), the pure Average force. In turn (0-14-0) and the ternary mixed force (4-4-5) finished joint third. Force (0-0-17) lost the competition with both the fewest wins, highest losses and the worst balance of average % survivors after each battle. In terms of force costing, (5-7-0) is slightly cheaper than (10-0-0) so represents best value for money within the competition rules.

3.2.2 Heterogeneous fighting strength (HFS) as an outcome predictor in MMW

The variations in force compositions serve as a test of Lanchester's HFS as a precombat battle outcome indicator in MMW. 26/42 battles went to the force with the highest precontact HFS, representing 62% of the battles of these 42 battles in the double round-robin competition 6 battles (2, 6, 19, 23, 27 & 40) resulted in matched HFS. These are impossible to predict by HFS alone before the battle takes place, so deducting these from the total, 26/36 went to the force with the highest precontact HFS, representing 72% of the battles.

Of those 10 battles that had the outcome reversed from the precombat ranking, 5 had precombat HFS that were within 1 of each other, so a small initial fluctuation in the survivors from 1st turn fire exchange could indeed reverse the precombat HFS ranking. The same argument also applied to the 6 battles with matched fighting strengths. Close inspection of these 16 battles suggests after the 1st turn fire exchange the outcome went to the side with the higher HFS. Only one battle (37) with (0-0-17) vs. (0-7-8) had matched attrition rates down until halfway through when (0-0-17) eventually dominated.

For losses between forces initially matched or close in fighting strength, a tipping point will quickly be reached as the number of units reduces by attrition, swinging the HFS to one side. As the number of troops reduces the chance of zero hits in the fire exchange significantly rises as shown in the tables in Appendix C. If we look at the 50% percentiles in the cumulative probability tables in Appendix C from the number of troops firing, these suggest the average number of hits we might expect in fire exchange. This number of hits can be less than the mean half the time, and more than this half the time. If we assume a 1st turn fire exchange where one force gets a low number of hits and the other a high number of hits, this corresponds to odds of ½ (force one low hits) x ½ (force two high hits) = ¼ of the time. Thus 10/42 battles is in the right proportion from this crude analysis for HFS reversing after the first fire exchange due to loss of troops.

3.2.3 1st salvo advantage and salvo intensity as an outcome predictor in MMW

We noted in our earlier paper that the side that inflicts the first major losses on their opponent gains an advantage in fighting strength, and is more likely to secure victory in the remainder of the fire exchange. However based purely on 1st salvo advantage, battles were won in 22/42 or 52% using this criterion. This suggests 1st salvo advantage is not as critical in winning battles in MMW compared to superior HFS. Across all battles the average % survivors for 1st salvo advantage was 35%, compared to 24% average % survivors for 2nd salvo reply, suggesting there is a slight advantage in firing first.

Appendix B introduced the concept of Salvo Intensity

$$\frac{[1 + P_b P_r - P_b]}{[1 - P_r]} \tag{8}$$

where P_r and P_b are the probabilities of hitting an opponent for red (1st salvo, subscript r) and blue (2nd salvo, subscript b) forces defined in Appendix B by taking the precombat average probabilities from the force composition based on their fractions in **Table D3**. From the analysis if the Salvo Intensity ≥1 the first salvo force is predicted to win, or if less than 1, the second salvo force. Using this as a battle outcome predictor gave the correct outcome 28/42 battles, representing 67% of the total battles.

3.2.4 Battle duration, fighting intensity and salvo intensity

Taylor [13] introduced the concept of 'Fighting Intensity', $\sqrt{P_r * P_b}$ Calculating the average 'Fighting Intensity' $\sqrt{P_r * P_b}$ as shown in **Table D3** and plotting this against the number of moves for the battle to end gives (**Figure 2**).

The critical value of Pearson's Correlation Coefficient at 95% confidence and 40 degrees of freedom (DoF) is 0.304, suggesting the trend is significant at the 95% confidence level. The average 'Fighting Intensity' is a guide for battle duration in MMW, with higher 'Fighting Intensity' i.e. improved odds in hitting reducing the number of moves to complete the battle.

Alternatively Appendix B introduced the concept of Salvo Intensity, Eq. (8). Taking the precombat average probabilities from the force composition based on their fractions from the forces in **Table D3**, calculating the average 'Salvo Intensity' and plotting this against the number of moves for the battle to end gives (**Figure 3**).

The critical value of Pearson's Correlation Coefficient at 99% confidence and 40 degrees of freedom (DoF) is 0.393, suggesting the trend is significant at the 99% confidence level.

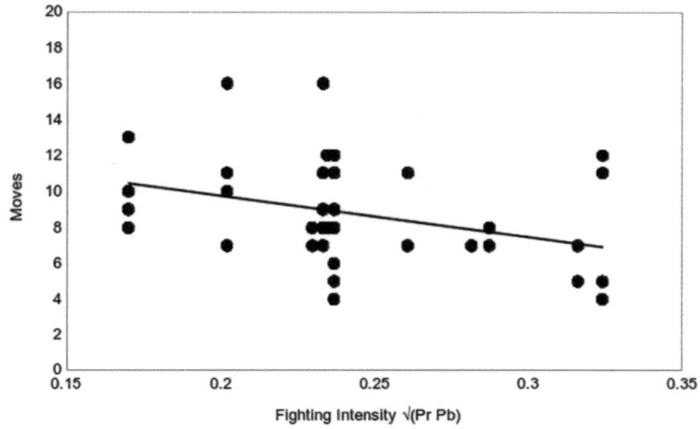

Figure 2.
*Moves for battle to end vs. average 'Fighting Intensity' from all runs in **Table D3**. The correlation coefficient for the data is R = −0.382 (40 DoF).*

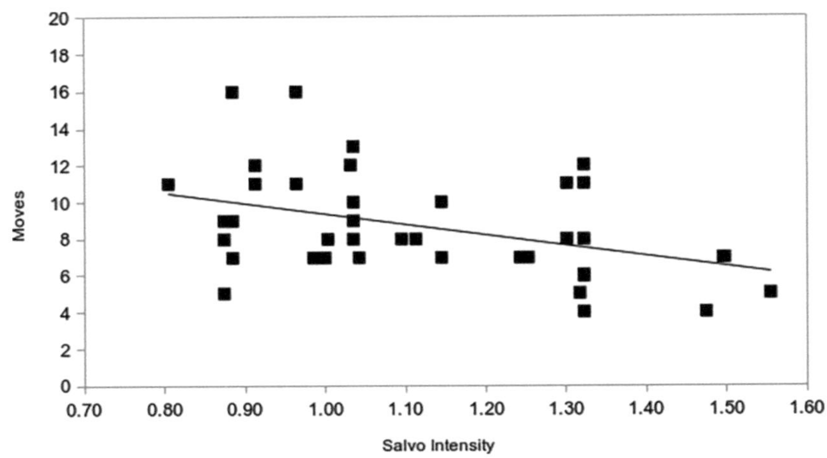

Figure 3.
Moves for battle to end vs. 'Salvo Intensity' from all runs in ***Table D3****. The correlation coefficient for the data is* $R = -0.408$ *(40 DoF).*

4. Conclusions

The precombat Lanchester's heterogenous fighting strength HFS appears to be a good indicator of battle outcomes in MMW and a strong guide to outcomes after the 1st turn fire exchange has occurred. Tracking this parameter turn by turn gives good guidance on eventual outcomes. This is despite the mismatch between the assumptions in Lanchestrian models (simultaneous, deterministic) and the simple manoeuvre and fire MMW (sequential turns & salvo firing with stochastic exchange of fire). Our salvo model and the concept of a 'Salvo Intensity' is also a good indicator for battle outcomes in MMW. It is also a better guide for battle duration in MMW, compared to Taylor's 'Fighting Intensity'. This highlights that improved odds in hitting enhances the chance of winning and reduces the number of moves to complete the battle. The probability for successful fire exchange in Appendix C suggests that the force with more troops that have higher probability of hitting (better quality in MMW) will achieve a higher rate of attrition, with fluctuations in hits due to dice rolling. This is a probabilistic explanation for the process described by Lanchester equations akin to Mangulis [14]. The Biblical encouragement for those face a foe with heterogeneous forces still stands.

> *When thou goest out to battle against thine enemies, and seest horses, and chariots, and a people more than thou, be not afraid of them.*
>
> *Deut 20:1*

A. Derivation of Lanchester's heterogeneous equations

For engagements of mixed blue and mixed red troops, we assume two things. First, that Lanchester's equations hold for each component of the forces, but second, they hold for the total number of troops. More specifically, if the total number of red troops is R and blue troops is B, then as Lanchester holds for the total number of troops:

$$\frac{dR}{dt} = -kB \tag{A1}$$

and:

$$\frac{dB}{dt} = -qR \tag{A2}$$

However, the red troops are made of mixed troops, which we label by i, so that R_i are the number of red troops of type i, and B_i are the number of blue troops of type i. Here, i might label the ability of the troops (Good, Average, or Poor) or else label the functions (infantry, cavalry/armour, artillery). As Lanchester holds for these components of the main force, we have:

$$\frac{dR_i}{dt} = -k_{ij}B_j \tag{A3}$$

And:

$$\frac{dB_i}{dt} = -q_{ij}R_j \tag{A4}$$

Where we follow the Einstein summation convention, that repeated indices are summed over. By definition:

$$R = \sum_i R_i \tag{A5}$$

And similarly for B.

$$B = \sum_i B_i \tag{A6}$$

Note that red and blue do not have to have the same number of component forces. Red might have good, average, and poor; blue might have only good and poor.

Because the Lanchester equations hold for R and B, we know that fighting strength is conserved, (Lanchester's square law).

That is to say,

$$-kB^2 + qR^2 = -kB(0)^2 + qR(0)^2 = const \tag{A7}$$

This conservation of fighting strength also allows for determination of the ratio of red troops to blue troops. From the chain rule, $\frac{d}{dt}\frac{R}{B} = \frac{\frac{dR}{dt}B - R\frac{dB}{dt}}{B^2}$From Eqs. (A1) and (A2) this becomes $\frac{d}{dt}\frac{R}{B} = \frac{qR^2-kB^2}{B^2}$. The numerator is the difference in fighting strengths which, from Eq. (A7), is constant. Hence as B^2 is positive, the ratio of red troops to blue troops increases if $\frac{qR^2-kB^2}{B^2} > 0$and decreases if $\frac{qR^2-kB^2}{B^2} < 0$. If the fighting strengths are equal, the ratio of red troops to blue troops remains constant throughout the battle.

Take Eq. (A3) and sum over i. Then:

$$\sum_i \frac{dR_i}{dt} = \frac{dR}{dt} = \sum_i - k_{ij}B_j \tag{A8}$$

Substitution from Eq. (A1) gives:

$$-kB = \sum_i - k_{ij}B_j \tag{A9}$$

And so:

$$-kB^2 = B\sum_i - k_{ij}B_j = \sum_m B_m \sum_i - k_{ij}B_j \tag{A10}$$

The right-hand side (RHS) of Eq. (A10) can be expanded and written as:

$$(k_{11}B_1 + k_{12}B_2 + \ldots k_{1n}B_n + k_{21}B_1 + k_{22}B_2 + \ldots k_{nn}B_n)(B_1 + B_2 + \ldots B_n) \tag{A11}$$

So:

$$RHS = (k_1B_1 + k_2B_2 + \ldots k_nB_n)(B_1 + B_2 + \ldots B_n) \tag{A12}$$

In which:

$$k_1 = k_{11} + k_{21} + \ldots k_{n1} etc \tag{A13}$$

This is the result for mixed troops given by Lepingwell [7].
Notice from Eq. (A9) that:

$$k = \left(\frac{1}{B}\right)\sum_i k_{ij}B_j \tag{A14}$$

That is to say, one would identify the k_{ij} to begin with, and then determine k from the above formula. As B and $B_i$have exponential time dependence, these will be eliminated, giving k as a constant.

B. Salvo model for warfare

Assume that the two troops take turns with salvo fire, at the same rate. We label the side with the 1st Salvo as Red, R, and the side with the 2nd Salvo as Blue, B. The basic model we assume turn by turn in fire exchange is:

$$R(n+1) = R(n) - kB(n+1) \tag{B1}$$

and

$$B(n+1) = B(n) - qR(n) \tag{B2}$$

Dividing (B1) by (B2):

$$\frac{R(n+1)}{B(n+1)} = \frac{[R(n) - kB(n+1)]}{[B(n) - qR(n)]} \tag{B3}$$

Or substitution for B(n + 1) from (B2):

$$\frac{R(n+1)}{B(n+1)} = \frac{[(1+kq)R(n) - kB(n)]}{[B(n) - qR(n)]} \tag{B4}$$

Dividing top and bottom by B(n):

$$\frac{R(n+1)}{B(n+1)} = \frac{\left[(1+kq)\frac{R(n)}{B(n)} - k\right]}{\left[1 - q\frac{R(n)}{B(n)}\right]} \tag{B5}$$

So that:

$$Z(n+1) = \frac{[(1+kq)Z(n) - k]}{[1 - qZ(n)]} \tag{B6}$$

This recurrence relation for Z(n + 1), the ratio of R(n+1) to B(n+1), is a first-order rational difference equation. As k and q are both real quantities, this is a Riccati recurrence relation for Z(n), which can be transformed into a linear equation.

Suppose now that Z(n) = 1 + p(n), where p(n) is greater than 1. That is to say, we believe that after the nth salvo, there are more red troops remaining than there are blue troops. In that case:

$$Z(n+1) = \frac{[1 + kq - k + kqp(n)]}{[1 - q - p(n)]} \tag{B7}$$

Now, the smallest value the numerator can have is when p(n) = 0, so the numerator is always bigger than $1 + kq - k$. The biggest denominator also occurs when p(n) is zero. Hence we know that:

$$Z(n+1) \geq \frac{[1 + kq - k]}{[1 - q]} \tag{B8}$$

If Z(n + 1) is also great than 1, then we have a condition that if red is greater than blue after n salvos, it remains so after n + 1 salvos.

Now for Z(n + 1) > 1, we require:

$$\frac{[1 + kq - k]}{[1 - q]} \geq 1 \tag{B9}$$

So if R(0) > B(0) and the above condition holds, then R will win.

We can define Good, Average and Poor component fractions in MMW by

$$\varphi_{G,A,P} = \frac{N_{G,A,P}}{N_G + N_A + N_P} \tag{B10}$$

$$\text{with } \varphi_G + \varphi_A + \varphi_P = 1 \tag{B11}$$

and the average probability for a force in MMW by the probability of hitting in the game multiplied by their fraction

$$P_{r,b} = P_G\varphi_G + P_A\varphi_A + P_P\varphi_P \tag{B12}$$

In terms of MMW we define (B9) as the 'Salvo Intensity'

$$\frac{[1 + P_bP_r - P_b]}{[1 - P_r]} \tag{B13}$$

C. Cumulative binomial distribution tables for 1D6 results

The cumulative binomial distribution [18] calculates the probability of getting up to a certain number of successes in a given number of trials, where each trial has only two possible outcomes (success or failure) and the probability of success is constant for each trial. It is described in Eq. (A12).

$$P(X \leqslant k) = \sum_{i=0}^{[k]} \frac{n!}{i!(n-i)!} p^i (1-p)^{n-i} \tag{C1}$$

where p is the probability of success in a Bernoulli trial, i is an integer from 0 to n, n is the number of trials and [k] is the greatest integer $\leq$ k.

This sums the probabilities of all possible outcomes from 0 successes up to the specified number of successes, giving 100% in total. In our case these relate to the probability of getting a certain number of specified 1D6 outcomes (such as a six) from repeated dice rolls. For a 1D6 being rolled, there is a 5/6 = 83.3% chance of not getting a six, and a 1/6 = 16.7% of getting a six, with the cumulative probability ending at 100%.

The following tables give the cumulative probabilities for 1D6 (1 to 20 consecutive rolls, in the columns) on getting either a 6, 5, 6 or 4–6 (the common dice outcomes used in MMW). The cumulative % chance for getting the number of dice roll results is shown in the rows. They are quoted to 1d.p. or 1/1000. Thus for example the chance of getting 10 sixes ($(1/6)^{10}$ = 1.65e-8) is excluded from the table, although it remains a faint probability. The first row in (**Tables C1–C3**) corresponds to the data presented in Appendix B in our earlier paper [2].

From **Table C1** on rolling 10D6 there is a 16.2% chance of not getting a single six, with the most likely outcomes ranging from 1 to 3 sixes, and a 0.2% chance of getting 6 sixes. To confirm this 500 runs of 10D6 were rolled using Google Dice. The expected cumulative probabilities in red (column 10 in **Table C1**) is plotted against the measured cumulative probabilities in green (**Figure C1**).

The Chi squared test revealed no difference between the expected and measured cumulative probabilities.

	1	2	3	4	5	6	7	8	9	10	11	12	13	14	15	16	17	18	19	20
0	83.3	69.4	57.9	48.2	40.2	33.5	27.9	23.3	19.4	16.2	13.5	11.2	9.3	7.8	6.5	5.4	4.5	3.8	3.1	2.6
1	100.0	97.2	92.6	86.8	80.4	73.7	67.0	60.5	54.3	48.5	43.1	38.1	33.6	29.6	26.0	22.7	19.8	17.3	15.0	13.0
2	—	100.0	99.5	98.4	96.5	93.8	90.4	86.5	82.2	77.5	72.7	67.7	62.8	57.9	53.2	48.7	44.4	40.3	36.4	32.9
3	—	—	100.0	99.9	99.7	99.1	98.2	96.9	95.2	93.0	90.4	87.5	84.2	80.6	76.8	72.9	68.9	64.8	60.7	56.7
4	—	—	—	100.0	100.0	99.9	99.8	99.5	99.1	98.5	97.5	96.4	94.9	93.1	91.0	88.7	86.0	83.2	80.1	76.9
5	—	—	—	—	—	100.0	100.0	100.0	99.9	99.8	99.5	99.2	98.7	98.1	97.3	96.2	95.0	93.5	91.8	89.8
6	—	—	—	—	—	—	—	—	100.0	100.0	99.9	99.9	99.8	99.6	99.3	99.0	98.5	97.9	97.2	96.3
7	—	—	—	—	—	—	—	—	—	—	100.0	100.0	100.0	99.9	99.9	99.8	99.7	99.5	99.2	98.9
8	—	—	—	—	—	—	—	—	—	—	—	—	—	—	100.0	100.0	99.9	99.9	99.8	99.7
9	—	—	—	—	—	—	—	—	—	—	—	—	—	—	—	—	100.0	100.0	100.0	99.9
10	—	—	—	—	—	—	—	—	—	—	—	—	—	—	—	—	—	—	—	100.0

Table C1.
Cumulative probability outcomes from the Binomial Distribution for repeated 1D6 rolls requiring a 6 to be a successful hit in the game.

	1	2	3	4	5	6	7	8	9	10	11	12	13	14	15	16	17	18	19	20
0	66.7	44.4	29.6	19.8	13.2	8.8	5.9	3.9	2.6	1.7	1.2	0.8	0.5	0.3	0.2	0.2	0.1	0.1	0.0	0.0
1	100.0	88.9	74.1	59.3	46.1	35.1	26.3	19.5	14.3	10.4	7.5	5.4	3.9	2.7	1.9	1.4	1.0	0.7	0.5	0.3
2	—	100.0	96.3	88.9	79.0	68.0	57.1	46.8	37.7	29.9	23.4	18.1	13.9	10.5	7.9	5.9	4.4	3.3	2.4	1.8
3	—	—	100.0	98.8	95.5	90.0	82.7	74.1	65.0	55.9	47.3	39.3	32.2	26.1	20.9	16.6	13.0	10.2	7.9	6.0
4	—	—	—	100.0	99.6	98.2	95.5	91.2	85.5	78.7	71.1	63.2	55.2	47.6	40.4	33.9	28.1	23.1	18.8	15.2
5	—	—	—	—	100.0	99.9	99.3	98.0	95.8	92.3	87.8	82.2	75.9	69.0	61.8	54.7	47.8	41.2	35.2	29.7
6	—	—	—	—	—	100.0	100.0	99.7	99.2	98.0	96.1	93.4	89.6	85.1	79.7	73.7	67.4	60.9	54.3	47.9
7	—	—	—	—	—	—	—	100.0	99.9	99.7	99.1	98.1	96.5	94.2	91.2	87.3	82.8	77.7	72.1	66.1
8	—	—	—	—	—	—	—	—	100.0	100.0	99.9	99.6	99.1	98.3	96.9	95.0	92.5	89.2	85.4	80.9
9	—	—	—	—	—	—	—	—	—	—	100.0	99.9	99.8	99.6	99.1	98.4	97.3	95.7	93.5	90.8
10	—	—	—	—	—	—	—	—	—	—	—	100.0	100.0	99.9	99.8	99.6	99.2	98.6	97.6	96.2
11	—	—	—	—	—	—	—	—	—	—	—	—	—	100.0	100.0	99.9	99.8	99.6	99.3	98.7
12	—	—	—	—	—	—	—	—	—	—	—	—	—	—	—	100.0	100.0	99.9	99.8	99.6
13	—	—	—	—	—	—	—	—	—	—	—	—	—	—	—	—	—	100.0	100.0	99.9
14	—	—	—	—	—	—	—	—	—	—	—	—	—	—	—	—	—	—	—	100.0

Table C2.
Cumulative probability outcomes from the binomial distribution for repeated 1D6 rolls requiring a 5, 6 to be a successful hit in the game.

	1	2	3	4	5	6	7	8	9	10	11	12	13	14	15	16	17	18	19	20
0	50.0	25.0	12.5	6.3	3.1	1.6	0.8	0.4	0.2	0.1	0.0	0.0	0.0	0.0	0.0	0.0	0.0	0.0	0.0	0.0
1	100.0	75.0	50.0	31.3	18.8	10.9	26.3	3.5	2.0	1.1	0.6	0.3	0.2	0.1	0.0	0.0	0.0	0.0	0.0	0.0
2	—	100.0	87.5	68.8	50.0	34.4	57.1	14.5	9.0	5.5	3.3	1.9	1.1	0.6	0.4	0.2	0.1	0.1	0.0	0.0
3	—	—	100.0	93.8	81.3	65.6	82.7	36.3	25.4	17.2	11.3	7.3	4.6	2.9	1.8	1.1	0.6	0.4	0.2	0.1
4	—	—	—	100.0	96.9	89.1	95.5	63.7	50.0	37.7	27.4	19.4	13.3	9.0	5.9	3.8	2.5	1.5	1.0	0.6
5	—	—	—	—	100.0	98.4	99.3	85.5	74.6	62.3	50.0	38.7	29.1	21.2	15.1	10.5	7.2	4.8	3.2	2.1
6	—	—	—	—	—	100.0	100.0	96.5	91.0	82.8	72.6	61.3	50.0	39.5	30.4	22.7	16.6	11.9	8.4	5.8
7	—	—	—	—	—	—	—	99.6	98.0	94.5	88.7	80.6	70.9	60.5	50.0	40.2	31.5	24.0	18.0	13.2
8	—	—	—	—	—	—	—	100.0	99.8	98.9	96.7	92.7	86.7	78.8	69.6	59.8	50.0	40.7	32.4	25.2
9	—	—	—	—	—	—	—	—	100.0	99.9	99.4	98.1	95.4	91.0	84.9	77.3	68.5	59.3	50.0	41.2
10	—	—	—	—	—	—	—	—	—	100.0	100.0	99.7	98.9	97.1	94.1	89.5	83.4	76.0	67.6	58.8
11	—	—	—	—	—	—	—	—	—	—	—	100.0	99.8	99.4	98.2	96.2	92.8	88.1	82.0	74.8
12	—	—	—	—	—	—	—	—	—	—	—	—	100.0	99.9	99.6	98.9	97.5	95.2	91.6	86.8
13	—	—	—	—	—	—	—	—	—	—	—	—	—	100.0	100.0	99.8	99.4	98.5	96.8	94.2
14	—	—	—	—	—	—	—	—	—	—	—	—	—	—	—	100.0	99.9	99.6	99.0	97.9
15	—	—	—	—	—	—	—	—	—	—	—	—	—	—	—	—	100.0	99.9	99.8	99.4
16	—	—	—	—	—	—	—	—	—	—	—	—	—	—	—	—	—	100.0	100.0	99.9
17	—	—	—	—	—	—	—	—	—	—	—	—	—	—	—	—	—	—	—	100.0

Table C3.
Cumulative probability outcomes from the binomial distribution for repeated 1D6 rolls requiring a 4, 5, 6 to be a successful hit in the game.

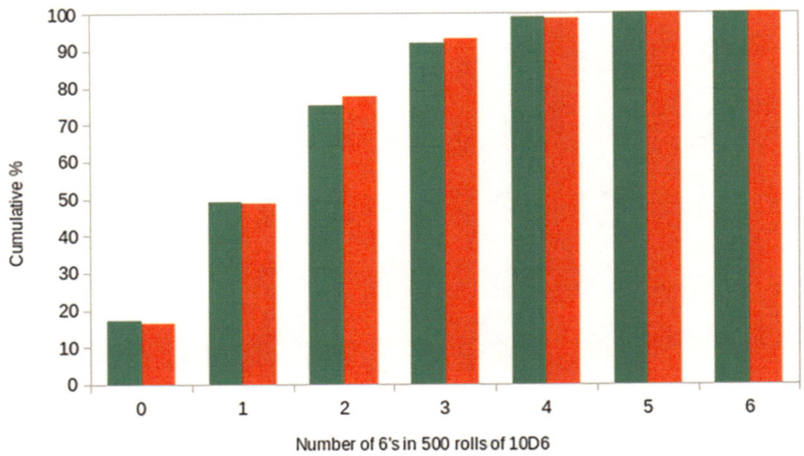

Figure C1.
Expected (red) vs. measured (green) cumulative probabilities for the number of sixes in 500 separate 10D6 rolls by Google Dice.

D. Table of results: Battle outcomes in the double round-robin tournament

Run	1st salvo G-A-P	2nd salvo G-A-P	Heterogeneous fighting strengths 1st salvo & 2nd salvo	Prediction	Outcome	% Survivors measured 1st salvo & 2nd salvo	Moves
1	10-0-0	5-7-0	33 < 34	5-7-0 Win	5-7-0 Win	0.0 & 57.1	7
2	10-0-0	0-14-0	33 ~ 33	Draw	10-0-0 Win	30.0 & 0.0	6
3	10-0-0	0-7-8	42 > 38	10-0-0 Win	10-0-0 Win	50.0 & 0.0	4
4	10-0-0	0-0-17	50 > 48	10-0-0 Win	10-0-0 Win	20.0 & 0.0	5
5	10-0-0	6-0-6	40 > 36	10-0-0 Win	10-0-0 Win	30.0 & 0.0	4
6	10-0-0	4-4-5	37 ~ 37	Draw	10-0-0 Win	40.0 & 0.0	5
7	5-7-0	0-14-0	34 > 33	5-7-0 Win	0-14-0 Win	0.0 & 42.8	10
8	5-7-0	0-7-8	48 > 38	5-7-0 Win	5-7-0 Win	60.0 & 0.0	7
9	5-7-0	0-0-17	58 > 48	5-7-0 Win	5-7-0 Win	75.0 & 0.0	7
10	5-7-0	6-0-6	44 > 36	5-7-0 Win	5-7-0 Win	57.1 & 0.0	7
11	5-7-0	4-4-5	40 > 37	5-7-0 Win	4-4-5 Win	0.0 & 61.5	8
12	0-14-0	0-7-8	49 > 38	0-14-0 Win	0-14-0 Win	57.1 & 0.0	9
13	0-14-0	0-0-17	65 > 48	0-14-0 Win	0-14-0 Win	71.4 & 0.0	8

Run	1st salvo G-A-P	2nd salvo G-A-P	Heterogeneous fighting strengths 1st salvo & 2nd salvo	Prediction	Outcome	% Survivors measured 1st salvo & 2nd salvo	Moves
14	0-14-0	6-0-6	49 > 36	0-14-0 Win	0-14-0 Win	82.3 & 0.0	8
15	0-14-0	4-4-5	42 > 37	0-14-0 Win	0-14-0 Win	78.5 & 0.0	7
16	0-7-8	0-0-17	55 > 48	0-7-8 Win	0-7-8 Win	13.3 & 0.0	13
17	0-7-8	6-0-6	37 > 36	0-7-8 Win	0-7-8 Win	86.7 & 0.0	5
18	0-7-8	4-4-5	38 > 37	0-7-8 Win	4-4-5 Win	0.0 & 17.7	16
19	0-0-17	6-0-6	48 ~ 48	Draw	6-0-6 Win	0.0 & 66.7	8
20	0-0-17	4-4-5	48 < 54	4-4-5 Win	4-4-5 Win	0.0 & 61.5	9
21	6-0-6	4-4-5	36 < 37	4-4-5 Win	4-4-5 Win	0.0 & 84.6	8

Table D1.
Predicted vs. measured battle outcomes in battles between forces with heterogeneous forces composed of Good, G, Average, A and Poor, P, troops. Heterogeneous fighting strengths (HFS) are calculated from Eq. (7).

Run	1st salvo G-A-P	2nd salvo G-A-P	Heterogeneous fighting strengths 1st salvo & 2nd salvo	Prediction	Outcome	% Survivors measured 1st salvo & 2nd salvo	Moves
22	5-7-0	10-0-0	34 > 33	5-7-0 Win	5-7-0 Win	33.3 & 0.0	7
23	0-14-0	10-0-0	33 ~ 33	Draw	0-14-0 Win	71.4 & 0.0	8
24	0-7-8	10-0-0	38 < 42	10-0-0 Win	10-0-0 Win	0.0 & 40.0	9
25	0-0-17	10-0-0	48 < 50	10-0-0 Win	10-0-0 Win	0.0 & 40.0	12
26	6-0-6	10-0-0	36 < 40	10-0-0 Win	10-0-0 Win	0.0 & 20.0	11
27	4-4-5	10-0-0	37 ~ 37	Draw	10-0-0 Win	0.0 & 40.0	7
28	0-14-0	5-7-0	33 < 34	5-7-0 Win	5-7-0 Win	0.0 & 41.6	11
29	0-7-8	5-7-0	38 < 48	5-7-0 Win	5-7-0 Win	0.0 & 8.3	16
30	0-0-17	5-7-0	48 < 58	5-7-0 Win	5-7-0 Win	0.0 & 58.3	11
31	6-0-6	5-7-0	36 < 44	5-7-0 Win	5-7-0 Win	0.0 & 75.0	8
32	4-4-5	5-7-0	37 < 40	5-7-0 Win	5-7-0 Win	0.0 & 66.7	7
33	0-7-8	0-14-0	38 < 49	0-14-0 Win	0-7-8 Win	53.3 & 0.0	13
34	0-0-17	0-14-0	48 < 65	0-14-0 Win	0-14-0 Win	0.0 & 50.0	9
35	6-0-6	0-14-0	36 < 49	0-14-0 Win	6-0-6 Win	50.0 & 0.0	12
36	4-4-5	0-14-0	37 < 42	0-14-0 Win	4-4-5 Win	23.1 & 0.0	8

Run	1st salvo G-A-P	2nd salvo G-A-P	Heterogeneous fighting strengths 1st salvo & 2nd salvo	Prediction	Outcome	% Survivors measured 1st salvo & 2nd salvo	Moves
37	0-0-17	0-7-8	48 < 55	0-7-8 Win	0-0-17 Win	47.1 & 0.0	10
38	6-0-6	0-7-8	36 < 37	0-7-8 Win	6-0-6 Win	75.0 & 0.0	8
39	4-4-5	0-7-8	37 < 38	0-7-8 Win	4-4-5 Win	84.6 & 0.0	8
40	6-0-6	0-0-17	48 ~ 48	Draw	6-0-6 Win	41.7 & 0.0	11
41	4-4-5	0-0-17	54 > 48	4-4-5 Win	4-4-5 Win	38.5 & 0.0	11
42	4-4-5	6-0-6	37 > 36	4-4-5 Win	6-0-6 Win	0.0 & 33.3	12

Table D2.
Predicted vs. measured battle outcomes in battles between forces with heterogeneous forces composed of Good, G, Average, A and Poor, P, troops. Heterogeneous fighting strengths (HFS) are calculated from Eq. (7). The order for first salvo fire is reversed from Table A.

D.1 Table of results: Battle outcomes in the double round-robin tournament

Run	1st salvo G-A-P 'Red'	P_r	2nd salvo G-A-P 'Blue'	P_b	$\sqrt{P_r . P_b}$ fighting intensity	$(1 + P_b P_r - P_b)/(1 - P_r)$ salvo intensity	Outcome (1st, 2nd salvo)	Moves
1	10-0-0	0.33	5-7-0	0.24	0.28	1.25	5-7-0 Win = 2nd	7
2	10-0-0	0.33	0-14-0	0.17	0.24	1.32	10-0-0 Win = 1st	6
3	10-0-0	0.33	0-7-8	0.17	0.24	1.32	10-0-0 Win = 1st	4
4	10-0-0	0.42	0-0-17	0.17	0.27	1.55	10-0-0 Win = 1st	5
5	10-0-0	0.42	6-0-6	0.25	0.32	1.47	10-0-0 Win = 1st	4
6	10-0-0	0.37	4-4-5	0.27	0.32	1.32	10-0-0 Win = 1st	5
7	5-7-0	0.24	0-14-0	0.17	0.20	1.15	0-14-0 Win = 2nd	10
8	5-7-0	0.24	0-7-8	0.17	0.20	1.15	5-7-0 Win = 1st	7
9	5-7-0	0.4	0-0-17	0.17	0.26	1.50	5-7-0 Win = 1st	7
10	5-7-0	0.33	6-0-6	0.25	0.29	1.24	5-7-0 Win = 1st	7
11	5-7-0	0.24	4-4-5	0.22	0.23	1.10	4-4-5 Win = 2nd	8
12	0-14-0	0.17	0-7-8	0.17	0.17	1.03	0-14-0 Win = 1st	9

Run	1st salvo G-A-P 'Red'	P_r	2nd salvo G-A-P 'Blue'	P_b	$\sqrt{P_r \cdot P_b}$ fighting intensity	$(1 + P_b P_r - P_b)/(1 - P_r)$ salvo intensity	Outcome (1st, 2nd salvo)	Moves
13	0-14-0	0.17	0-0-17	0.17	0.17	1.03	0-14-0 Win = 1st	8
14	0-14-0	0.17	6-0-6	0.33	0.24	0.87	0-14-0 Win = 1st	8
15	0-14-0	0.17	4-4-5	0.32	0.23	0.88	0-14-0 Win = 1st	7
16	0-7-8	0.17	0-0-17	0.17	0.17	1.03	0-7-8 Win = 1st	13
17	0-7-8	0.17	6-0-6	0.33	0.24	0.87	0-7-8 Win = 1st	5
18	0-7-8	0.17	4-4-5	0.32	0.23	0.88	4-4-5 Win = 2nd	16
19	0-0-17	0.17	6-0-6	0.33	0.24	0.87	6-0-6 Win = 2nd	8
20	0-0-17	0.17	4-4-5	0.32	0.23	0.88	4-4-5 Win = 2nd	9
21	6-0-6	0.25	4-4-5	0.22	0.23	1.11	4-4-5 Win = 2nd	8
22	5-7-0	0.24	10-0-0	0.33	0.28	0.99	5-7-0 Win = 1st	7
23	0-14-0	0.17	10-0-0	0.33	0.24	0.87	0-14-0 Win = 1st	8
24	0-7-8	0.17	10-0-0	0.33	0.24	0.87	10-0-0 Win = 2nd	9
25	0-0-17	0.25	10-0-0	0.42	0.32	0.91	10-0-0 Win = 2nd	12
26	6-0-6	0.25	10-0-0	0.42	0.32	0.91	10-0-0 Win = 2nd	11
27	4-4-5	0.27	10-0-0	0.37	0.32	1.00	10-0-0 Win = 2nd	7
28	0-14-0	0.17	5-7-0	0.24	0.20	0.96	5-7-0 Win = 2nd	11
29	0-7-8	0.17	5-7-0	0.24	0.20	0.96	5-7-0 Win = 2nd	16
30	0-0-17	0.17	5-7-0	0.4	0.26	0.80	5-7-0 Win = 2nd	11
31	6-0-6	0.25	5-7-0	0.33	0.29	1.00	5-7-0 Win = 2nd	8
32	4-4-5	0.22	5-7-0	0.24	0.23	1.04	5-7-0 Win = 2nd	7
33	0-7-8	0.17	0-14-0	0.17	0.17	1.03	0-7-8 Win = 1st	13
34	0-0-17	0.17	0-14-0	0.17	0.17	1.03	0-14-0 Win = 2nd	9

Run	1st salvo G-A-P 'Red'	P_r	2nd salvo G-A-P 'Blue'	P_b	$\sqrt{P_r \cdot P_b}$ fighting intensity	$(1 + P_b P_r - P_b)/(1 - P_r)$ salvo intensity	Outcome (1st, 2nd salvo)	Moves
35	6-0-6	0.33	0-14-0	0.17	0.24	1.32	6-0-6 Win = 1st	12
36	4-4-5	0.32	0-14-0	0.17	0.23	1.30	4-4-5 Win = 1st	8
37	0-0-17	0.17	0-7-8	0.17	0.17	1.03	0-0-17 Win = 1st	10
38	6-0-6	0.33	0-7-8	0.17	0.24	1.32	6-0-6 Win = 1st	8
39	4-4-5	0.32	0-7-8	0.17	0.23	1.30	4-4-5 Win = 1st	8
40	6-0-6	0.33	0-0-17	0.17	0.24	1.32	6-0-6 Win = 1st	11
41	4-4-5	0.32	0-0-17	0.17	0.23	1.30	4-4-5 Win = 1st	11
42	4-4-5	0.22	6-0-6	0.25	0.23	1.03	6-0-6 Win = 2nd	12

Table D3.
Probabilities for hitting for 1st salvo, Pr and 2nd salvo based on the force compositions using Eq. (8). The 'Fighting Intensity' and the 'Salvo Intensity 'are calculated and the outcomes and moves to complete the battle are shown.

Author details

Mark Flanagan[1], David Lambert[2], Trevor C. Lipscombe[3], Adrian Northey[2] and Ian M. Robinson[2]*

1 NHS Business Service Authority, Newcastle Upon Tyne, UK

2 Hearts of Oak, North Yorks, UK

3 Catholic University of America, Washington, DC, USA

*Address all correspondence to: a.anser4u@gmail.com

References

[1] Lanchester FW. The principle of concentration: N- Square law. In: Chapter 5, Aircraft in Warfare: The Dawn of the Fourth Arm. London: Constable; 1916. Available from: https://tinyurl.com/3h9j63xv

[2] Flanagan M, Lambert D, Lipscombe TC, Northey A, Robinson IM. Lanchester's fighting strength as a Battle outcome predictor applied to a simple fire and manoeuvre Wargame. In: Recent Advances in Monte Carlo Methods. London: IntechOpen; 2024. ISBN 978-1-83769-362-7. Available from: https://www.intechopen.com/chapters/1169739

[3] Lanchester's "Fighting Strength" as a battle outcome predictor applied to a simple fire and manoeuvre wargame – supporting GIFS. Available from: https://tinyurl.com/25zve5dt

[4] Historical Analysis for Defence and Security Symposium 2024. Available from: https://tinyurl.com/47mkesf4

[5] 'Möbius Mayhem 2123 '—'Vernichtungsschlacht' in the vacuum of space', Wargames Vault. The Game is Free. 2023. Available from: https://tinyurl.com/mu7bp4en

[6] Lanchester FW. The N- Square law in its application. In: Chapter 7, Aircraft in Warfare: The Dawn of the Fourth Arm. London: Constable; 1916. Available from: https://tinyurl.com/4henp7u2

[7] Lepingwell JWR. The Laws of combat? Lanchester reexamined. International Security. 1987;**12**(1): 89-134. Available from: https://www.jstor.org/stable/2538918

[8] Colegrave RK, J. M. Hyde 'the Lanchester square-law model extended to a (2,2) conflict'. IMA Journal of Applied Mathematics. 1993;**51**: 95-109. DOI: 10.1093/imamat/51.2.95

[9] MacKay NJ. Lanchester models for mixed forces with semi-dynamical target allocation. Journal of the Operational Research Society. 2009; **60**(10):1421-1427. Available from: https://www.jstor.org/stable/40295701

[10] Lin KY, MacKay NJ. The optimal policy for the one-against-many heterogeneous Lanchester model. Operations Research Letters. 2014;**42**(6–7):473-477. DOI: 10.1016/j.orl.2014.08.008

[11] Armstrong MJ. A stochastic salvo model for naval surface combat. Operations Research. 2005;**53**(5): 830-841. Available from: https://www.jstor.org/stable/25146917

[12] Armstrong MJ. The salvo combat model with a sequential exchange of fire. Journal of the Operational Research Society. 2014;**65**(10):1593-1601. Available from: https://www.jstor.org/stable/24505020

[13] Taylor JG. Lanchester-Type Models of Warfare, Volume II. 1980. Available from: https://apps.dtic.mil/sti/tr/pdf/ADA090843.pdf

[14] Mangulis V. Lanchester-type equations of combat which account for multiple killing hits. Operations Research. 1980;**28**(3):560-569. Available from: http://www.jstor.com/stable/170404

[15] Kim D, Moon H, Park D, Shin H. An efficient approximate solution for

stochastic Lanchester models. Journal of the Operational Research Society. 2017; **68**(11):1470-1481. Available from: https://www.jstor.org/stable/10.2307/48759954

[16] Lanchester's Heterogenous Fighting Strength (HFS) for MMW.xslx. 2024. Available from: https://wp.me/p7dM0-1ks

[17] Binomial Distribution. Distribution. 2024. Available from: https://en.wikipedia.org/wiki/Binomial_distribution

[18] Cornell JA. Experiments with Mixtures Designs, Models, and the Analysis of Mixture Data. J. New York: Wiley & Sons; 1990. ISBN: 9780471522218

[19] Online GIFS of Games in This Paper. 2024. Available from: https://tinyurl.com/52nvaye9

Section 2

Applications of Game Theory in Strategic Decision-Making

Chapter 3

Application and Analytical Workflows of Evolutionary Games in Assisted Analytical Decision Making at the Planning Stage: An Example of Urban Renewal and Rural Tourism Planning

Youmei Zhou, Hao Lei, Yanzhao Bi, Jianlin Zhu, Jingjing Chen and Dandan Wang

Abstract

Evolutionary game theory (EGT) offers a valuable framework for analyzing complex decision-making processes in the early stages of urban renewal and rural tourism planning. This chapter employs EGT to explore its application and analytical workflows through two case studies. The process involves identifying stakeholders, constructing game models, simulating the game process, and analyzing the outcomes to identify optimal strategies. This approach provides insights into the dynamics of the decision-making process, facilitates scenario exploration, and supports evidence-based policy development. The chapter also discusses the importance of considering assumptions and data limitations, drawing on the positive response received from an academic journal publication. The goal is to refine and summarize the findings in higher dimensions, expanding the application of EGT in planning and design.

Keywords: evolutionary game theory, decision dynamics, stakeholders, simulation, sustainability

1. Introduction

1.1 Background of urban renewal and rural tourism planning

Urban renewal and rural tourism planning are key components of sustainable regional development strategies [1, 2]. As urban areas continue to grow and evolve, there is a pressing need to address the challenges of aging infrastructure, declining neighborhoods, and social inequality. Urban renewal serves as a strategic effort to revitalize decayed or underutilized urban spaces, improve living conditions, and foster economic development. By contrast, rural areas face different challenges, including population decline, lack of infrastructure, and the erosion of traditional ways of life. Rural tourism planning has emerged to rejuvenate these areas by leveraging cultural heritage, natural resources, and unique landscapes to attract visitors, which in turn fosters local economic development.

Urban renewal typically involves large-scale projects focusing on infrastructure improvements, housing, transportation, and public spaces [3, 4]. The aim is to create vibrant, livable urban environments that can adapt to the needs of modern societies. On the other hand, rural tourism is often built around preserving the natural environment and local traditions while providing opportunities for sustainable economic development [5]. In both contexts, planning is a complex process involving multiple stakeholders, ranging from government agencies to local communities and private investors. The challenge is not only to achieve economic growth but also to ensure that it is inclusive, sustainable, and respectful of local cultural identities.

1.2 Significance of evolutionary game theory in decision-making processes

Evolutionary Game Theory (EGT) offers a powerful framework for analyzing decision-making in complex systems like urban renewal and rural tourism planning. Unlike traditional game theory, which assumes that players have perfect information and make fully rational decisions, EGT accommodates the reality of bounded rationality and gradual adaptation. This makes it particularly suited for situations where stakeholders have diverse interests and objectives and must adapt their strategies over time based on the actions of others.

In the context of urban renewal and rural tourism planning, EGT provides a way to model the interactions between different stakeholders, such as local governments, developers, residents, and tourists. Each stakeholder group may have different preferences and constraints, leading to a dynamic decision-making environment where strategies evolve over time.

By applying EGT, planners and policymakers can simulate the potential outcomes of various strategies and interactions between stakeholders. This allows for the identification of stable strategies. Such simulations can provide valuable insights into the long-term consequences of different policy choices and help identify strategies that are more likely to lead to desirable outcomes for all stakeholders.

1.3 Objective of the chapter

1. To Apply Evolutionary Game Theory to analyze decision-making processes in the early stages of urban renewal and rural tourism planning.

2. To provide insights into the dynamics of the decision-making process and offer practical guidance for policymakers and planners.

3. To explore and summarize the findings and expand the application and analysis workflow of EGT in planning and design.

2. Evolutionary game theory (EGT) review

Being a further development of game theory, evolutionary game theory predicts the strategy choice of each subject, simulates the game process between subjects to develop the optimal strategy to maximize the group's interests and achieve the Nash equilibrium, and finally forms the corresponding auxiliary decision-making recommendations [6]. Evolutionary Game Theory (EGT) is an extension of classical game theory, which contains basic principles of evolutionary biology. Paying attention to the change of strategies of the whole population rather than just individuals, EGT analyzed how strategies maximize the group's benefit.

Taking advantage of the simulation of full-volume data for assisted decision-making, evolutionary game theory, and deep learning is crucial to the community regeneration field, which needs the active participation of effective and speedy quantitative models [7]. Instead of assuming decision-makers are entirely rational or that released information is perfect, the core value of EGT is to explain how strategic behaviors make the population adapt to change. Therefore, EGT believes humans arrive at a dynamic balance via a long-time trial and error behavior. Evolutionary Game Theory is widely used in the study of social institutions, behavioral norms, and other socio-economic issues due to the fact that the dynamic choice mechanism it uses can be applied to multiple games of the system and form an evolutionary and stable strategy [8].

2.1 Advantages of EGT in analyzing complex decision-making process

The advantages of EGT in Analyzing complex decision-making processes are numerous when it comes to its dynamic and adaptive function. Firstly, EGT allows the modeling of evolving environments and behaviors, reflecting how strategies change over time with dynamic conditions. In the context of the rural tourist development decision-making process, concerning the long-term dynamic procedure such as stakeholder interactions, alternative strategies for addressing these rural tourism development decision-making problems have been proposed from different perspectives, including game theory [9]. What is more, unlike classic game theory, EGT fully understands bounded rationality and adaptation, which means that the non-terminated change of strategy composition is essential. It offers permissions for adjustment to those agents in reality, generally through trials but not calculations. Different from TGT players, who are rational and forward-looking, EGT players are used to a relatively gradual and myopic process [10].

Also, advantages like revealing internal dependence between stakeholders, zooming in on the effect of one agent's strategy changing on the other agencies, and showing how regional work may lead to a national outcome cannot be ignored. Beyond that, evolutionary game theory is famous for dynamic simulation and its interdisciplinary application in analyzing complex decision-making processes. On the one hand, the results of some complex systems are not linear, making them hard to predict; however, EGT helps to explain some phenomena with dynamic simulations. The latter part of the book will provide EGT dynamic models that help explain the section. On the other hand, EGT allows interdisciplinary application. It means that EGT is widely applicable in different fields, such as economics, biology, political science, and social science, which shows its inclusive utilization. Therefore, decision-makers are allowed to apply EGT as an effective method to plan different tasks.

2.2 Previous applications of EGT in planning and design

The example of EGT being used in city planning is not rare: In 2024, Luan and his team built up an evolutionary game model for the construction of the MESB in the

Bohai Sea, with the most effective factors that affect the related decision of whether nearby companies will participate being analyzed. With the tool of replication dynamics analysis and tripartite game model consisting of marine enterprises, local governments, and the public, the evolutionary equilibrium conditions and evolutionary stabilization strategies of the three parties using MATLAB R2020a are analyzed and constructed [11]. Similarly, in order to explore the dynamics between local governments and the public in the context of EIA, Liang and his team constructed an evolutionary game model to portray the interests of the two players through times of simulation and sensitivity analysis, which in the end accomplished their goals [12].

3. Case study 1: Urban renewal planning in the context of community age-appropriate renewal practice in Baoshan District, Shanghai

In the decision-making process of urban renewal and rural tourism planning, stakeholder identification is crucial [6]. In the first case study, age-friendly renewal is used as a scenario for evolutionary game simulation, and modeling simulation is carried out through stakeholder's interest analysis to back-project the governance strategy to verify the effectiveness of the proposed system model in complex social problems.

3.1 Identification of stakeholders

This case study identifies three primary stakeholders in the community regeneration context:

The first stakeholder is government: This includes administrative bodies responsible for policymaking and oversight. The second is enterprises: Primarily private developers, construction firms, and other social organizations that provide services or invest in regeneration projects. The last is Residents: Local populations affected by urban renewal efforts, ranging from tenants to indigenous populations in communities.

Each stakeholder has distinct interests and goals, which evolve based on their interactions with one another. The importance of stakeholder identification is emphasized because the success of regeneration projects relies on cooperation between these groups. The interaction between stakeholders is dynamic, with changing strategies influenced by both personal benefits and societal needs [8, 13].

3.2 Construction of game models

In Evolutionary Game Theory (EGT), constructing a game model requires an overview of how stakeholders interact within an urban renewal framework. The case study builds a three-way evolutionary game model with the government, businesses and residents as game parties. The aim of the game is to simulate the strategic interactions between them:

The government's strategy can vary between active governance (providing subsidies, facilitating participation) and loose governance; firms must choose between active participation (full involvement in the project) and passive participation (failing to contribute sufficiently); and residents can either actively participate in community redevelopment or passively observe and adapt to changes.

The strategies chosen by these participants are not static but evolve over time. As shown in **Tables 1–3**, stakeholders in urban renewal and rural tourism planning can choose different strategies based on their objectives and interactions with other stakeholders [6]. By analyzing their payoffs - subsidies, penalties, cost savings or reputation enhancement - the game model can predict the behavior of each stakeholder in various scenarios of urban renewal. The model aims to reflect the complex social, economic and political dynamics of community renewal projects.

3.3 Simulation of the game process

Once a game model has been constructed, simulating interactions between stakeholders allows potential outcomes to be explored. In this case, the simulation takes into account the following evolutionary nature of the game:

Stakeholders adjust their strategies based on previous outcomes to maximize returns over time; government intervention in terms of subsidies or penalties plays an important role in influencing the behavior of firms and residents; and the simulation incorporates real-world constraints, such as funding limitations and the involvement of local organizations (**Figure 1**).

Parameter	Meaning	Remarks
S	Subsidies given to businesses and residents when the government is aggressive in governance	
C_e	Costs paid by the government when it adopts a proactive strategy	
M	Penalties for negative participation of enterprises when the government is active in governance	
α	The percentage of cost saved by the government and residents' income increased when community planners are involved in the project	$\alpha \in [0, 1]$
β	The proportion of cost savings for the government and enterprises when the government and enterprises cooperate	$\beta \in [0, 1]$
P_1	Direct and indirect benefits such as economic and prestige gained by government departments when community governance is progressed	
P_2	Loss of prestige due to poor living experience of residents when the government is passive	$P_1 > P_2$
C_1	Costs to businesses when they actively participate	
C_2	The cost of active participation by the government and passive participation by enterprises	$C_1 > C_2$
R_f	The basic benefits of public services provided by market players	
R_e	Additional economic benefits for each market player due to increased reputation in the industry	
R_i	Indirect benefits to residents in terms of physical, mental, and quality of life improvement	
L_p	Compensation costs for residents who suffer as a result of the negative participation strategy of market players	
L_n	The benefit loss of residents when the community renewal effect is poor	

Table 1.
Parameter setting table of the three-party game.

Strategy selection			Earnings		
Government	Companies	Residents	Government	Companies	Residents
C	C	C	$P_1 - S - (1 - \alpha - \beta)C_e$	$S + R_f + R_e - C_1$	$\alpha R_i - R_f$
C	C	D	$P_1 - S - (1 - \beta)C_e$	$S + R_e - C_1$	R_i
C	D	C	$P_1 - S + M - (1 - \alpha)C_e$	$S - M + R_f - C_2 - L_p$	$\alpha R_i + L_p - R_f - L_n$
C	D	D	$M - S - C_e$	$S - M - C_2$	$-L_n$
D	C	C	$-C_e - P_2$	$R_f + R_e - C_1$	$R_i - R_f$
D	C	D	$-C_e$	$-C_1$	$-L_n$
D	D	C	$-C_e - P_2$	$R_f - L_p - C_2$	$-L_n - R_f$
D	D	D	$-C_e$	$-C_2$	$-L_n$

Table 2.
The matrix of benefits of the three party game in case study 1.

Equilibrium point	Eigenvalues of the Jacobian matrix			Symbol of the real part	Stability conclusion
	λ_1	λ_2	λ_3		
$E_1(0,0,0)$	$C_2 - C_1$	$M - S$	$-R_f$	$(-,-,-)$	ESS
$E_2(1,0,0)$	$S - \mathrm{M}$	$L_n - R_f + \alpha R_i$	$C_2 - C_1 + M + R_e$	$(\mathrm{X},0,+)$	Instability point
$E_3(0,1,0)$	$C_1 - C_2$	$P_1 - S - \beta C_e$	$L_n - R_f + R_i$	$(+,-,+)$	Instability point
$E_4(0,0,1)$	R_f	$C_2 - C_1 + M + R_e$	$M + P_1 + P_2 - S + \alpha C_e$	$(+,+,+)$	Instability point
$E_5(1,1,0)$	$S - P_1 + \beta C_e$	$C_1 - C_2 - M - R_e$	$L_n - L_P - R_f - (1-\alpha)R_i$	$(-,-,-)$	ESS
$E_6(1,0,1)$	$R_f - L_n - \alpha R_i$	$C_2 - C_1 + L_P + M + R_e$	$S - P_1 - P_2 - M - \alpha C_e$	$(-,+,-)$	Instability point
$E_7(0,1,1)$	$R_f - L_n - R_i$	$C_1 - C_2 - M - R_e$	$P_1 + P_2 - S + \alpha C_e - \beta C_e$	$(-,-,+)$	Instability point
$E_8(1,1,1)$	$L_P - L_n + R_f + (1-\alpha)R_\mathrm{i}$	$S - P_1 - P_2 - \alpha C_e + \beta C_e$	$C_1 - C_2 - L_P - M - R_e$	$(-,-,-)$	Instability point
$E_9(0,y_1,z_1)$	0	0	a_1	$(0,0,+)$	Instability point
$E_{11}(x_1,0,z_2)$	0	a_2	a_3	$(0,-,-)$	Uncertain point
$E_{12}(x_2,y_2,0)$	a_4	a_5	a_6	$(\mathrm{X},\mathrm{X},\mathrm{X})$	Uncertain point

Table 3.
The parameters matrix of benefits of the three-party game.

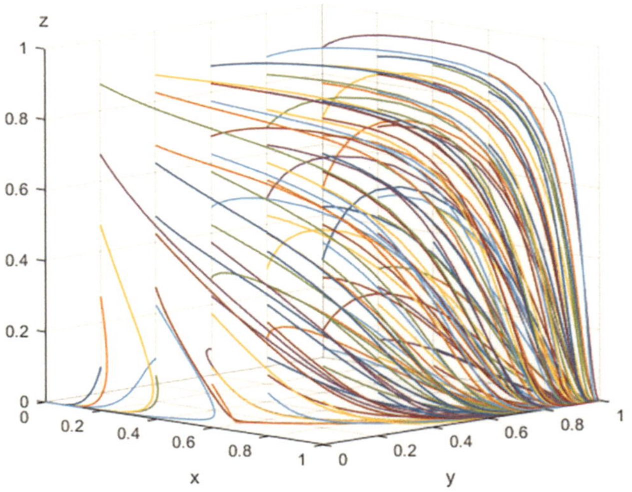

Figure 1.
The result of 50 evolutions by initial values.

The simulation game process reveals potential scenarios in which stakeholders either cooperate for successful regeneration or fail to coordinate their interests, leading to suboptimal outcomes. Through multiple iterations of the simulation, planners can predict the impact of different decisions (e.g., changes in policy or resource allocation) on the long-term sustainability of the project.

3.4 Analysis of outcomes and optimal strategies

The aim of analyzing the results of the modeling is to identify the best strategy. In the case of community regeneration, the goal is to find a balance where all stakeholders benefit:

For government, active governance (providing subsidies and working with businesses and residents) usually leads to better outcomes, but this has to be balanced against the financial costs of subsidies; for businesses, they are more likely to be active when they are given financial incentives and the penalties for non-participation are higher. However, over-reliance on subsidies can stifle innovation; for residents, participation increases when they perceive a tangible improvement in their quality of life, which can be achieved through better services and active community engagement.

By comparing the outcomes across different simulations, the study identifies evolutionarily stable strategies—scenarios where no stakeholder has an incentive to change their strategy because it leads to the most favorable long-term results. These strategies help urban planners and policymakers make informed, evidence-based decisions that maximize stakeholder engagement and ensure the long-term success of urban renewal projects.

4. Case study 2: Rural tourism planning

In rural tourism development, the identification of stakeholders is essential to ensure sustainable growth and avoid conflict. They have different roles and incentives, which change as the tourism landscape evolves.

4.1 Identification of stakeholders

The case is set in the context of rural tourism in China, which can be divided into three categories of stakeholders: government agencies, which play a crucial role by setting regulations, defining strategic directions, and providing incentives for development. They are also responsible for meeting the needs of local communities and protecting natural and cultural resources. In China, for example, multiple levels of government are responsible for planning and coordinating rural tourism activities [14]; the second category is tourism enterprises, ranging from small local businesses to large corporations, which are usually driven by economic efficiency [15]. These entities invest in infrastructure, provide services, and market rural tourism destinations to attract tourists. Their involvement is often driven by short- or long-term profit potential. The final category, residents, are equally important stakeholders as they are directly affected by tourism development. They may actively participate, oppose, or remain neutral. Residents' support or opposition can influence the success of a tourism project, so their participation is crucial for sustainable rural tourism [16].

4.2 Construction of game models

Once stakeholders are identified, the next step is to construct game models that represent the interactions between them. In the context of evolutionary game theory (EGT), these models simulate the decision-making process of each stakeholder, considering their respective strategies and payoffs. The goal is to predict how each party will behave in response to the actions of others.

In rural tourism, there are different strategies available for each party: government, business and residents. For governments, they can choose active governance (providing financial subsidies, policy support, and enforcing regulations) or passive governance (minimal intervention, less supervision). Active governance tends to lead to better development outcomes but at a higher cost. For businesses, tourism firms can choose to invest for the long term, maintaining and developing tourism assets in a sustainable manner, or they can choose short-term development, focusing on profit maximization with little regard for sustainability. For residents, residents can actively participate in tourism development (providing services and preserving cultural heritage) or they can passively oppose it, which can undermine or slow down projects.

The game model accounts for the interdependence of these strategies. For example, a government that offers generous subsidies may encourage businesses to pursue long-term investments, while residents may be more likely to participate if they see tangible benefits. Conversely, insufficient government support may lead to short-term strategies by enterprises and limited resident involvement. The model simulates how stakeholders' choices evolve over time as they react to the strategies of others (**Tables 4** and 5).

4.3 Simulation of the game process

After constructing the game model, it is necessary to simulate the interactions between stakeholders over time. This process provides insights into how different strategies will play out in the real world.

The simulation involves running multiple iterations of the game, with each stakeholder adjusting their strategy based on previous outcomes. For example, if residents initially choose to oppose tourism development but later see that participation leads to

Game subject	Parameter	Parameter definition	Remarks
Government	C_f	Financial allocations spent by the government to improve rural infrastructure	
	R_i	Financial revenue received by the government through taxation, etc.	
	R_j	Government's performance gains and enrichment image	
	C_t	Good policy environment, financial investment, special debts, bank loans, project declaration, etc., created by the government for the development of tourism scenic spots by enterprises	
	M_t	Financial penalties imposed by the government on tourism enterprises due to the loss of residents' interests and the destruction of scenic spots in the course of their operation	
	M_g	Financial compensation for rural residents due to excessive impact on their livelihood	
	M_{up}	Penalties imposed by higher government departments on relevant departments due to failure to meet set development targets	
	β	Proportion of missing compensation to residents when there is collusion between government and enterprises	
	δ	Unjustified gains made as a result of collusion between government and enterprises	
Tourism Enterprises	R_s	Proceeds from short-term tourism projects	$R_s > C_d$
	R_l	Long-term tourism benefits	$R_l > C_p$
	R_v	The image and branding of the tourism company in the industry, etc.	
	R_p	Facilitation of various aspects of tourism opening by the government as a result of active and effective cooperation with the government	$C_t > R_p$
	C_d	Costs incurred for participation in the development of tourist attractions	
	M_l	Compensation to residents of the village for excessive negative impact on their lives and the environment	
	$\propto$	The proportion of dividends paid to the village residents	
	C_p	Maintenance costs during the subsequent operation of the tourist attraction	
Village Residents	R_r	Financial compensation for the inhabitants during the initial phase of the tourism development	
	M_c	If the residents choose to actively participate in the opening of the tourist attraction, the cultural impact will lead to cultural progress in the village; if they choose to participate passively, the cultural impact will lead to a decline in the residents' life experience	
	R_k	Increased employment as well as increased economic returns to the village after the opening of the tourist attraction (when negative, meaning a brain drain from the village)	
	C_r	Decreased life experience for residents when passively participating in the development of a tourist attraction	

Game subject	Parameter	Parameter definition	Remarks
	M_v	Impacts on residents' lives and places of residence during the development and subsequent operation of tourist attractions	
	R_h	Residents' voice in the development of tourist attractions	

Table 4.
Tripartite game subject parameter settings.

increased income and community improvements, they may shift their strategy toward active engagement. Similarly, if enterprises notice that long-term investments yield better returns due to government subsidies and resident cooperation, they may focus more on sustainable development.

The game process also allows for the exploration of various scenarios, such as changes in government policy, shifts in market conditions, or environmental challenges. By modeling these different possibilities, the simulation can help predict potential conflicts or synergies between stakeholders. It can also highlight the conditions under which cooperation is most likely to occur, thus informing policy decisions (**Figure 2**).

For example, in China's rural tourism, government-led initiatives often set the tone for tourism development. The level of support provided to businesses and residents can determine the success or failure of a project. The simulation can help assess the impact of different government actions, such as offering more significant financial incentives to tourism enterprises or enhancing compensation for residents whose livelihoods are affected by tourism.

4.4 Analysis of outcomes and optimal strategies

Analyzing the results of the simulations is a way of finding a balance where all parties can benefit in the long term and identifying the best strategy for all stakeholders. The results of the simulations show that an active governance approach with government subsidies and support leads to better outcomes for all stakeholders. However, governments must balance the costs of intervention with active governance, as excessive subsidies can strain public resources. Long-term investment in tourism development often proves to be the most sustainable strategy for businesses. By maintaining and developing rural tourism assets, businesses can build brands and create lasting economic impact. However, this requires patience and a willingness to accept lower short-term profits. Active participation in tourism development can bring significant benefits to residents, such as increased income and employment opportunities. However, this depends on how well governments and businesses engage local residents and address their concerns.

5. Application and analytical workflows of EGT

5.1 Step-by-step guide to applying EGT in planning

5.1.1 Stakeholder identification

The first step in applying EGT in planning is to identify the stakeholders in the system that are modeled. In the model, every single stakeholder adapts its own

Strategy choice			Benefits		
Government	**Enterprises**	**Residents**	**Government**	**Enterprises**	**Residents**
C	C	C	$-C_f + R_i - C_t + R_j$	$(1-\alpha)R_l + R_v + R_p - C_d - C_P$	$\alpha R_l + R_r + M_c + R_k + R_h - M_v$
C	C	D	$-C_f + R_i - C_t - R_j - M_{up} + \delta$	$C_t + R_l - R_v - C_d - C_P + \delta$	$(1-\beta)R_r - M_c - C_r - M_v$
C	D	C	$-C_f + R_i - C_t + R_j + M_t - M_g - M_{up}$	$R_s - M_l - C_d - M_t - R_v$	$R_r + M_c + R_k + R_h + M_g + M_l$
C	D	D	$-C_f - C_t + R_j + M_t - M_g - M_{up}$	$R_s - M_l - C_d - M_t - R_v$	$R_r - M_c - R_k - C_r + M_g + M_l$
D	C	C	$-C_f + R_i$	$(1-\alpha)R_l + R_v + R_p - C_d - C_P$	$R_r + M_c + R_k + R_h + \alpha R_l - M_v$
D	C	D	$-C_f + R_i - R_j + \delta$	$R_l + R_v - C_d - C_P + \delta$	$(1-\beta)R_r - M_c - C_r - M_v$
D	D	C	$-C_f + R_i - R_j - M_{up}$	$R_s - R_v - C_d$	$M_c + R_k$
D	D	D	$-C_f - M_{up}$	$R_s - R_v - C_d$	$-M_c - C_r$

Table 5.
The matrix of benefits of the three-party game in case study 2

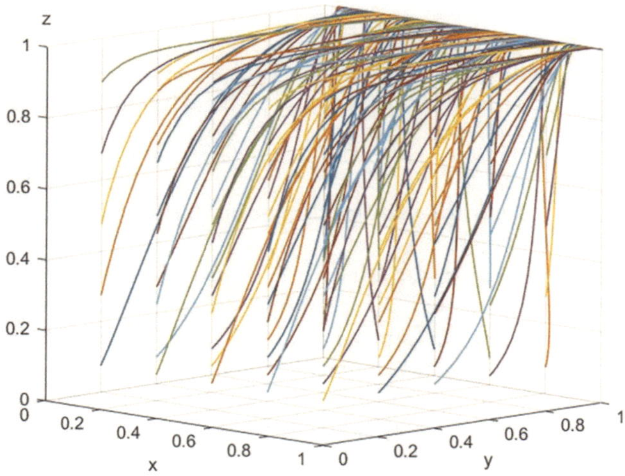

Figure 2.
Evolutionary path diagram of the 50 times evolutionary game of the three parties.

available strategies that influence the system's dynamics. For example, in an urban renewal and rural tourism scenario, there will be several different stakeholders, including local governments hoping to promote local economic growth, local business organizations seeking to attract more tourists, and residents aiming at reducing the loss of a comfortable living environment. Even though these stakeholders have different motivations, their benefits are tightly connected with local resources.

5.1.2 Game model construction

When stakeholder identification is finished, the next step will be constructing the game model. Each stakeholder sets up effective strategies and determines the gain or loss of each strategy. The stakeholders may choose among a range of strategies, such as investing in tourism advertisement, environmental conservation, or adapting local urban development projects. For example, the local government may choose local development projects that boost local economics, while local businesses prefer investing in tourism advertisement, which brings them more customers. Residents apply to environmental conservation, which increases living quality.

5.1.3 Game process simulation

There is an internal relationship running between stakeholders in the game process simulation, and the strategies evolve over time based on stakeholders' situations. Planners will be able to observe and later research how different strategies influence the whole system over time, and there may be collaborations, competitions, and conflicts among stakeholders. For example, in the rural tourist project, the simulation might model the investment in tourist infrastructure, which is supported by the local business. If excessive development leads to environmental degradation, the residents' goals will not be satisfied, which leads to a conflict between local businesses and residents.

5.1.4 Outcome analysis and strategy identification

The result of the simulation is supposed to be analyzed in order to define whether to choose stable strategies or recurring dynamics. By analyzing outcomes, long-term stable strategies can be determined.

5.2 Benefits of EGT in decision-making dynamics and scenario exploration

EGT showed its particular superiority in decision-making dynamics and scenario exploration. With its dynamic strategic planning, adaptive decision-making, and stable equilibrium identification, EGT provides the decision-making process of community regeneration with strengthened scientific support.

For example, if we regard each cycle of the chart below as one trial, then EGTs will get the corresponding result of every single decision in trials. The chart below will separate the scenario into two parts: Statement and Action. Firstly, the process begins with the environment in the Statement. Environment collects extensive perceptible data. Every time environmental conditions change, the agent will develop a new strategy for decision-making, such as situation and case classification, gaming behavior identification, and training and optimization of evolutionary game-supported decision models utilizing neural network algorithms and reinforcement learning. Then, after the agent changes a strategy, action comes (**Figure 3**).

The action part offers four types of community governance to choose from, including Government-Oriented, Market-Oriented, Co-production Participation, and Planner Participatory Community Renew, which are the strategy-based answers. Every time the extensive perceptible data, which is the beginning point, changes, one corresponding agent will change, and the model will output a new result.

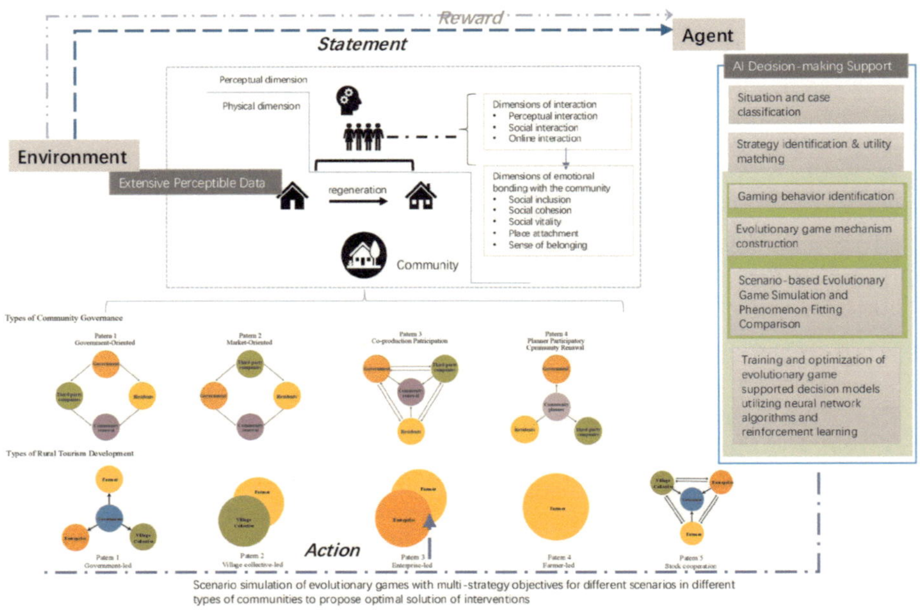

Figure 3.
The conceptual framework of workflow in urban and rural planning with reinforcement learning.

6. Considerations and limitations

6.1 Importance of assumptions in EGT

Evolutionary Game Theory relies on several assumptions to model decision-making processes. Understanding these assumptions is crucial for interpreting the results and applying EGT effectively. Key assumptions include:

- Bounded rationality: Decision-makers are not fully rational and make decisions based on limited information and cognitive abilities.
- Strategic interaction: The actions of one decision-maker affect the payoffs and strategies of others.
- Adaptation: Decision-makers learn from their experiences and adjust their strategies over time.

It is essential to critically assess the validity of these assumptions in the specific context of urban renewal and rural tourism planning. Deviations from these assumptions can lead to inaccurate predictions and suboptimal decision-making.

6.2 Data limitations and their impacts on analysis

Data availability and quality are critical factors in applying EGT. Limitations in data can arise from various sources:

- Incomplete data: Lack of information on stakeholders, strategies, and payoffs.
- Biased data: Data collected from specific sources may not represent the entire population.
- Dynamic data: Data changes over time, making it challenging to capture the evolving dynamics of the system.

Data limitations can impact the accuracy and reliability of EGT models. It is crucial to acknowledge these limitations and employ appropriate techniques to address them, such as data imputation, sensitivity analysis, and robustness checks.

6.3 Strategies for addressing assumption and data challenges

To address the challenges posed by assumptions and data limitations, several strategies can be employed:

- Validation and verification: Test the model against real-world data and expert knowledge to ensure its validity.
- Sensitivity analysis: Assess the impact of changes in assumptions and data on the model's outcomes.
- Robustness checks: Investigate the stability of the model under different scenarios and assumptions.

- Data collection and improvement: Improve data collection methods and expand the scope of data to capture a more comprehensive view of the system.

7. Academic journal publication and positive response

7.1 Overview of the published article

This chapter is based on a published academic article titled "Application and Analytical Workflows of Evolutionary Games in Assisted Analytical Decision Making at the Planning Stage - An Example of Urban Renewal and Rural Tourism Planning". The article presents a framework for applying EGT to analyze decision-making processes in urban renewal and rural tourism planning. It provides insights into the dynamics of the decision-making process and offers practical guidance for policymakers and planners.

7.2 Key insights and contributions of the article

The article makes several key contributions:

- *EGT framework for planning*: It establishes a systematic framework for applying EGT to analyze complex decision-making processes in planning.

- *Case studies*: The article demonstrates the application of EGT through two case studies: urban renewal and rural tourism planning.

- *Insights into dynamics*: It provides insights into the dynamics of stakeholder interactions and the evolution of strategies over time.

- *Guidance for policymakers*: The article offers practical guidance for policymakers and planners to make evidence-based decisions.

7.3 Implications for future research and practice

The article has several implications for future research and practice:

- *Expanding EGT applications*: Further research can explore the application of EGT to other planning domains, such as transportation, infrastructure, and environmental planning.

- *Model refinement*: Researchers can refine EGT models by incorporating more realistic assumptions and data sources.

- *Decision support systems*: EGT can be integrated into decision support systems to assist planners and policymakers in making informed decisions.

- *Policy development*: EGT can inform the development of evidence-based policies that promote sustainable and equitable development.

8. Refinement and summary of findings

In summary, this chapter uses the game method to simulate the complex interaction patterns of multi-stakeholders in rural tourism under different situations, and summarizes the specific patterns of effective cooperation among different stakeholders. The significance of this study lies in the use of the game method to predict the coping strategies and outcomes of each subject in different contexts, and to elaborate a win-win situation of cooperation among different stakeholders with the government actively taking the lead, enterprises developing in the long term, and residents actively participating. The model helps to promote multi-party cooperation in rural tourism and achieve sustainable development. Through the application of EGT, the model helps to promote multi-stakeholder cooperation in rural tourism and ultimately realizes sustainable development in urban and rural areas.

On the basis of affirming the importance of stakeholder cooperation and participation, the research in this chapter further innovatively establishes a strategic model of effective stakeholder cooperation, which further complements and improves the research on the complex relationships between stakeholders in tourism. In addition, this chapter innovatively uses the game method to clarify the complex relationship model between multi-stakeholders, deepening the understanding of multi-stakeholder relationships in the process of rural tourism development, and further deepening the research on the relationship between relevant stakeholders in the field of tourism.

This chapter innovatively uses the game approach to simulate the interaction between different stakeholders in rural tourism development, which provides useful and concrete practical guidance for promoting the sustainable development of rural tourism in the future. This innovative approach can inspire future research, provide a basis for decision-making, and validate the findings using artificial intelligence models.

9. Conclusion

The aim of this chapter is to construct a three-party evolutionary game model to analyze the strategic choices of government departments, tourism enterprises, and rural residents in the Chinese context, and to derive optimization strategies for sustainable development. Based on the simulation design, the core model of the evolutionary game functional module is initially established, and the strategy output and decision support paths that can be applied in the future digital twin community governance are developed to further interface with the strategy support platforms for city operation, urban management, and community management.

Through the application of EGT, an autonomous decision-making system based on reinforcement learning is proposed to provide optimal strategies for the multi-dimensional intersection of sociology with traditional spatial planning and governmental governance, and parameter setting and simulation are conducted with a specific community as a case study to validate the gaming behaviors and results, which is of great significance in solving the autonomous optimization problems in the future related city operation and urban planning. Meanwhile, the research in this chapter is based on computational social science and evolutionary game theory, investigates the social governance state evolution and weighted decision-making model, constructs a data and knowledge fusion-driven simulation method for the evolution of complex systems of social governance, proposes a systematic countermeasure for ubiquitous perception and ubiquitous participation of residents in the network era, and corroborates the importance of the social and perceptual dimensions of the digital twins in the network era.

Future research can be based on long-term multi-temporal and spatial cross-sectional data collection for model extraction, large-scale data collection, and the formation of a rural revitalization industry database in the future, so as to provide a scenario-deepening basis and a basis of judgment for the research direction of this topic based on AI model-assisted decision-making.

Acknowledgements

This research was funded by the National Natural Science Foundation of China (52208079).

Author details

Youmei Zhou[1*], Hao Lei[2], Yanzhao Bi[3], Jianlin Zhu[4], Jingjing Chen[5] and Dandan Wang[6]

1 Department of Environmental Design, School of Art and Design, Beijing Forestry University, Beijing, China

2 Department of Landscape Architecture, School of Architecture, Tsinghua University, Haidian, Beijing, China

3 Department of Landscape Architecture, The University of Sheffield, Western Bank, Sheffield, UK

4 Department of Economics, Collage of Liberal Arts and Sciences, University of Illinois at Urbana-Champaign, Washington, DC, US

5 School of Economic, Fudan University, Shanghai, China

6 Jilin Huayu Film and Television Culture Media Co., Ltd, Changchun, China

*Address all correspondence to: youmeizhou@bjfu.edu.cn

References

[1] Maculan LS, Dal Moro L. Strategies for inclusive urban renewal. In: Filho WL et al., editors. Sustainable Cities and Communities. Cham: Springer International Publishing; 2020. pp. 662-672

[2] Wanner A. The use of experiential acceptance in psychotherapy with emerging adults [Doctoral dissertation]. Vienna: University of Natural Resources and Life Science; 2022

[3] Huang L et al. Paths and strategies for sustainable urban renewal at the neighbourhood level: A framework for decision-making. Sustainable Cities and Society. 2020;**55**:102074

[4] Lin S-H et al. Evaluating the sustainability of urban renewal projects based on a model of hybrid multiple-attribute decision-making. Land Use Policy. 2021;**108**:105570

[5] Matiku SM. A Tourism Framework for Community-Driven Sustainable Livelihood: A Comparative Study. [Doctoral dissertation]. South Africa: University of Venda; 2021.

[6] Zhou Y et al. Using the dual concept of evolutionary game and reinforcement learning in support of decision-making process of community regeneration—Case study in Shanghai. Buildings. 2023; **13**(1):175

[7] Blokhuis E et al. Conflicts and cooperation in brownfield redevelopment projects: Application of conjoint analysis and game theory to model strategic decision making. Journal of Urban Planning and Development. 2012;**138**(3):195-205

[8] Shan H, Yang J. Sustainability of photovoltaic poverty alleviation in China: An evolutionary game between stakeholders. Energy. 2019;**181**:264-280

[9] Todd L, Leask A, Ensor J. Understanding primary stakeholders' multiple roles in hallmark event tourism management. Tourism Management. 2017;**59**:494-509

[10] Izquierdo LR, Izquierdo SS, Sandholm WH. An introduction to ABED: Agent-based simulation of evolutionary game dynamics. Games and Economic Behavior. 2019;**118**:434-462

[11] Luan D, Cao H, Qu T. An analysis of the tripartite evolutionary game for building the ecological security barrier in the Bohai Sea under the threshold of land-sea integration. Frontiers in Marine Science. 2024;**10**:1226665

[12] Liang Y et al. Promoting public participation in NIMBY facilities' EIA in urban planning: An evolutionary game model. Journal of Urban Planning and Development. 2024;**150**(1):04023064

[13] Shan S-N et al. Evolutionary game analysis of stakeholder behavior strategies in 'not in my backyard' conflicts: Effect of the intervention by environmental non-governmental organizations. Sustainable Production and Consumption. 2021;**28**:829-847

[14] Chen X, Huang J. Government intervention, diversification and firm performance. Management World. 2007; **1**(92):7

[15] Wang C, Xu H. Government intervention in investment by Chinese listed companies that have diversified into tourism. Tourism Management. 2011;**32**(6):1371-1380

[16] Huo T et al. Residents' participation in rural tourism and interpersonal trust in tourists: The mediating role of residents' perceptions of tourism impacts. Journal of Hospitality and Tourism Management. 2023;**54**:457-471

Chapter 4

Game Theory and Strategic Decision-Making in the Dark Souls Games

Tibor Guzsvinecz and Judit Szűcs

Abstract

This chapter examines the application of game theory to the Dark Souls games, which are critically acclaimed action role-playing games known for their challenging gameplay. Using game theory as an analytical framework, in this study, we use important concepts such as sequential decision-making, iterative learning, risk management, information asymmetry, and meta-learning to assess the Dark Souls games. In these games, players face high-stake strategic choices from combat encounters and boss fights to resource management and multiplayer interactions. By analyzing the Dark Souls games with concepts from game theory, this chapter can provide an understanding of the strategic decisions that define the game. This also allows us to understand how players navigate complex, high-risk scenarios in gaming, which may be applied to real-world decision-making as well.

Keywords: Dark Souls, game theory, iterative learning, risk management, sequential decision-making

1. Introduction

The Dark Souls games influenced game design and affected the gaming industry: more and more games are becoming similar to them [1]. Dark Souls challenges traditional game narratives by using environmental storytelling and minimalistic dialog. They also encourage players to explore and piece together the lore themselves. Their unforgiving difficulty level and "trial and error" gameplay create a unique learning curve [2]. This also mirrors real-life problem-solving and resilience-building. Additionally, the series addresses philosophical themes such as existentialism, struggle, and the cyclic nature of life and death. This can provide a ground for academic discussion in areas like ethics, narrative theory, and even mental health. It is possible to understand player experience and video games in general by examining these games [3–6]. Consequently, analyzing the Dark Souls games can help in the understanding of game design, storytelling, and player psychology. In this study, we analyze these games with the help of concepts from game theory.

Game theory offers a framework for analyzing strategic decision-making in situations where the outcome of one player's choice depends on the actions of others.

 IntechOpen

Initially, it was developed to model competitive interactions in economics and political science. Since then, game theory gained applications across disciplines such as biology [7], psychology [8], and video game design [9]. In gaming, the strategic nature of decision-making often mirrors real-world applications of game theory. This makes it an important tool for understanding player behavior and game dynamics.

Using game theory, we can explore situations in which players must make decisions while considering the potential strategies of others. In video games, these decisions can be categorized into player versus environment (PvE), player versus player (PvP), or cooperative multiplayer interactions. In the latter, players collaborate toward shared objectives. Games with a high level of difficulty like Dark Souls can offer a case study to understand how players navigate risks, manage resources, and adapt to evolving challenges.

Numerous other video games incorporate strategic decision-making in ways that can be analyzed using game theory. However, what sets Dark Souls apart from many of them is its unique combination of punishing difficulty, asymmetric multiplayer, and iterative gameplay. Players must learn from each death, refine their strategies, and overcome invaders (other players) and enemies controlled by artificial intelligence (AI) while managing limited resources. These mechanics reflect various aspects of game theory: decision-making under uncertainty, payoff structures, and mixed strategies.

This book chapter is structured as follows. Section 2 presents a short summary of the Dark Souls games. Section 3 talks about the payoff structures in them. Section 4 details the combat mechanics of these video games from a game theory perspective. Similarly, the multiplayer mechanics are assessed in Section 5. Next, information asymmetry, uncertainty, and meta-learning are detailed in Section 6. Lastly, conclusions are drawn in Section 7.

2. The Dark Souls games

The Dark Souls series, developed by FromSoftware, is an action role-playing game series known for its challenging gameplay, complex world design, and deep lore [10]. It is set in a dark, medieval fantasy universe where players control an undead character. Usually, this undead character goes through a so-called hero's journey, which is a concept used in fantasy and fiction. These games do not provide a helping hand for the players; thus, they have to learn everything about these games for themselves [11].

At its core, the Dark Souls series is about exploration, survival, and overcoming challenges. The game is played from a third-person perspective and players must navigate through interconnected areas filled with hostile enemies, powerful bosses, and environmental hazards. Players can customize their characters by choosing from various classes, such as a knight, mage, or thief (among others). They can also upgrade their weapons and armor at blacksmiths. These choices affect the players' combat style and how they interact with the game world. These are important as such games include a punishing difficulty that rewards patience and skill, a unique combat system that emphasizes timing and strategy, and an interconnected world that encourages exploration [12–14]. A screenshot from Dark Souls III can be observed in **Figure 1**.

Combat is considered one of the most important aspects of the Dark Souls games. However, the basics of the combat have to be introduced here to fully understand the remaining parts of the chapter. In the first two games, players only have a health and a stamina bar. In the third game, a mana bar is introduced as well. These three

Figure 1.
A screenshot from Dark Souls III: an enemy is running toward the player character.

bars can be seen in the upper left corner in **Figure 1**. Health is the red bar, mana is the blue bar, and stamina is the green bar. Naturally, health can be decreased by getting hit, while it can be replenished by using health potions or resting at bonfires (checkpoints). Stamina, on the other hand, can be decreased by running, dodging, or blocking attacks and by simply attacking enemies. The stronger the attack, the more stamina is drained. Stamina is replenished by not doing any of the above. Basically, the player character has to remain on standby (although walking is permitted). Mana is decreased by casting spells. Similarly to health, it can be replenished by using mana potions or resting at bonfires.

From a game-theoretic perspective, one of the most important characteristics of the Dark Souls games is their iterative nature: players frequently die; then, they respawn and try again with new knowledge. This aligns with the repeated games concept where players encounter the same or similar situations multiple times while they have to adjust their strategies based on previous outcomes. Each death provides the players with more information about enemy patterns, level layout, or optimal item use. This allows them to refine their approach in the following attempts. In game theory, this iterative learning process is similar to evolutionary games. There, strategies evolve based on their effectiveness in previous iterations.

Scientific studies have explored the role of iterative learning in gaming. They showed that players adapt their strategies based on repeated exposure to similar challenges. For example, a study by Vorderer et al. found that players in digital games undergo a process of "cognitive adaptation" [15]. According to this process, they adjust their strategies in response to changing game dynamics. This concept is highly applicable to Dark Souls as players repeatedly encounter challenging bosses or level sections. They gradually develop an optimal strategy through trial and error. Additionally, this adaptive process is enhanced in multiplayer settings. In Dark Souls, players can invade each other's worlds for PvP combat, which can create asymmetric encounters that further challenge strategic thinking. Asymmetric interactions often require players to adopt mixed strategies and dynamic decision-making to gain an advantage. This is particularly relevant in Dark Souls' invasion system since invaders usually face resource disadvantages and must rely on clever tactics or deception to defeat the host player.

Another key aspect of game theory relevant to Dark Souls is risk management. For example, a player may choose to explore a hidden path in Dark Souls. They know that while the path could contain valuable items or shortcuts, it may also lead to death. The player must weigh the potential rewards against the risk of losing in-game currency or progress. This is similar to how economists model decision-making under risk in real-world scenarios. Players' risk tolerance might be influenced by factors such as familiarity with the game, prior experiences, and the perceived difficulty of the challenge. In the Dark Souls games, players can exhibit a higher tolerance for risk after repeated failures since they grow more familiar with enemy patterns and level design. This can reduce the perceived uncertainty.

3. Payoff structures in the Dark Souls games

In game theory, a payoff structure defines the outcomes or rewards that players receive based on their choices and the circumstances they face. This is an important aspect of the Dark Souls experience. From the very beginning of the game, players are thrown into a world where every decision carries potential consequences, and progression is often built on the balance of risk management and strategic foresight. Payoffs take many forms in the Dark Souls games: souls (the game's currency), health, time, and even emotional satisfaction from defeating a tough enemy. However, these rewards are tempered by the ever-present risk of death. The death of the player's character results in the loss of collected souls and resets the player to the nearest checkpoint. Every choice the players make can be understood as a strategic decision that seeks to maximize their payoff while minimizing the associated risks.

3.1 The bonfire checkpoint system

Bonfires serve as checkpoints in the game. Near them, players can heal and replenish their healing items. However, resting at a bonfire comes with a trade-off: it respawns all (defeated) non-boss enemies in the game. This process is illustrated in **Figure 2**. Overall, this decision can be analyzed using a basic utility function, where the player evaluates the potential benefit (surviving longer) versus the cost (re-engaging respawned enemies). In some cases, the optimal strategy may be to rest frequently, mainly if the player is carrying a large number of souls and risks losing them upon death. However, in other scenarios, pushing forward may yield a greater payoff. There is a possibility that the player is close to discovering a new shortcut or defeating a boss. In economic terms, this is similar to making decisions based on opportunity cost, which is the cost of choosing one course of action over another. In Dark Souls, this trade-off is most evident in the tension between the need to progress and the desire to maintain safety.

3.2 Managing currency and experience

Souls not only serve as in-game currency but also as experience points in Dark Souls. They are a finite resource that players collect by defeating enemies and can be used to level up or purchase items. However, if the player's character dies, all of

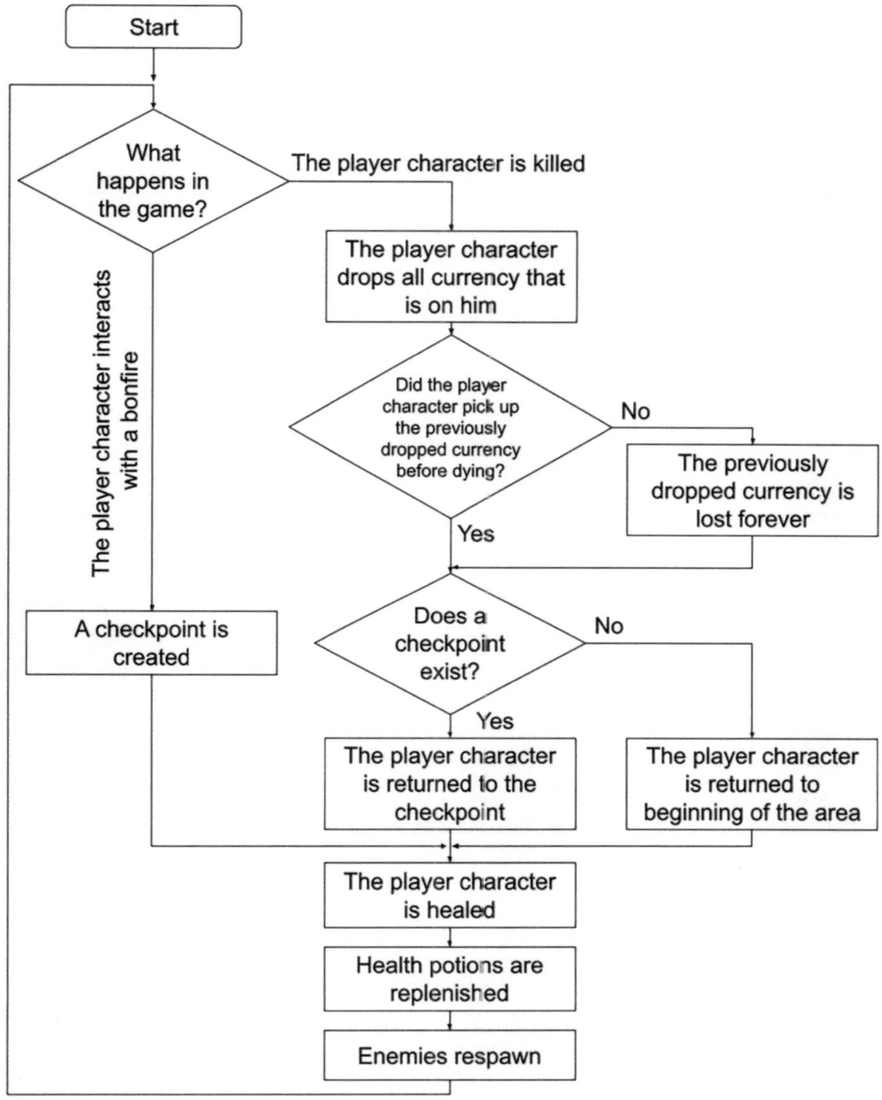

Figure 2.
The bonfire checkpoint system.

their collected souls are lost at the place of their death. There is only one opportunity to reclaim them by reaching the same spot without dying again. If they die before reaching that point, their souls are lost forever. This creates a risk-reward structure where the players must decide how far to push their luck before spending their souls or returning to safety. In game theory, this scenario can be described as a dynamic game: the players' strategy evolves based on their current situation. This includes how many souls they have collected, how far they are from a bonfire, and the strength of nearby enemies.

This dynamic can be likened to investment theory, where players must decide when to "cash in" their souls (i.e., return to the bonfire) versus continuing to invest (i.e., defeating more enemies for more souls). In essence, the player must manage risk aversion, which is a well-documented concept in game theory and economics. Risk-averse players may return to bonfires more frequently to bank their souls and make smaller, incremental gains. On the contrary, risk-seeking players may push forward while hoping for a bigger reward at the cost of greater danger.

Kahneman and Tversky suggest that players' risk preferences can change based on how they frame their potential gains and losses [16]. In the context of Dark Souls, players who have just gained a large number of souls may become more risk-averse, fearing the loss of a substantial reward. Conversely, players with few souls may be more willing to take risks, as their potential loss is minimal compared to the potential gain.

3.3 Health management

One of the core mechanics in Dark Souls is the management of the healing items, called Estus Flasks. They can be used to restore health in the game. Players start with a limited number of Estus Flasks, but they can be replenished at bonfires. Deciding when to use these flasks introduces another layer of strategic decision-making: players must balance the need for health against respawned enemies and the possibility of encountering even more dangerous ones later on. In game-theoretical terms, this can be viewed as a resource management problem. Each Estus Flask is a finite resource and the players' optimal strategy depends on their current state (i.e., how much health they have left) and the expected difficulty of upcoming encounters. This decision-making process is related to decision theory, mainly under conditions of uncertainty. As players progress through the game and become more familiar with enemy patterns and level layouts, their ability to make optimal decisions improves. This mirrors the concept of Bayesian learning as players change their strategies based on the information they gather through gameplay.

3.4 Risks in boss fights

Boss fights in Dark Souls represent some of the highest-stakes encounters in the game. These battles often feature enemies with large health pools, powerful attacks, and complex patterns. This forces the player to engage in extended combat where every mistake can be fatal. The risk-reward structure in boss fights is especially stark:

- *Risk:* The player risks death. This may result in the loss of a large number of souls and the need to retrace their steps to the boss arena.

- *Reward:* Defeating a boss yields large rewards: a huge quantity of souls, new items, and progress to the next area of the game.

From a game theory perspective, these encounters are high-stakes games where the players' optimal strategy involves careful observation of the boss's attack patterns and efficient resource management. Boss fights also introduce a level of uncertainty, mainly during the first few attempts since the player is still learning the boss's behaviors. This creates a situation where players must make decisions with incomplete information and they have to refine their strategy as they gather more data. The decisions of players can be framed as a payoff matrix where each decision (attack, defend, or heal) has an associated set of risks and rewards. The players' goal is to choose the option that maximizes their overall payoff. They have to balance the benefits of hitting the boss or healing with the potential long-term costs of making a mistake.

4. Combat in the Dark Souls games

Combat in Dark Souls is widely regarded as one of the most challenging and rewarding aspects of the game. The combat system is built on precision, timing, and resource management [17]. Each combat encounter, even against a low-level enemy, demands strategic decision-making and risk assessment. Combat can be modeled as a dynamic interaction where players must weigh the potential gains (defeating an enemy, progressing in the game, and earning souls) against the potential losses (taking damage, losing health, and dying). This creates a high-stakes environment where players are constantly balancing risks and rewards while making real-time decisions. Combat also aligns with game-theoretical concepts such as minimax theory, sequential decision-making, and adaptive strategies.

4.1 Optimizing combat strategies with the minimax theory

The minimax theory is often applied to two-player games like chess. The goal of minimax theory is to minimize the possible loss for a worst-case scenario while maximizing the potential gain. In Dark Souls, the players' objective is to minimize the damage they take (minimizing losses) while maximizing the damage they deal to enemies (maximizing gains). Each enemy presents unique challenges such as different attack patterns, movement speed, and damage type. However, unlike turn-based games, combat in Dark Souls occurs in real time. This adds another layer of complexity.

Consider a typical combat encounter: the player has to choose between offensive actions (e.g., attacking) and defensive actions (e.g., dodging and blocking). Each offensive action drains stamina, which is a finite resource that is also required for defensive actions like dodging. Therefore, overcommitting to an attack can leave the player vulnerable to enemy strikes. This means that the players must balance their offensive and defensive moves to ensure they maximize damage output while minimizing the risk of being hit. The complexity of the combat increases with imperfect information since the players do not know what attack an enemy will use next. This can force them to make decisions based on incomplete data.

4.2 Finite-horizon sequential games

In game theory, sequential games are scenarios where players take turns making decisions and each decision influences the next. In the case of finite-horizon sequential games, players have a limited number of opportunities to act. Still, while Dark Souls is not turn-based in a traditional sense, its combat system can be analyzed as a series of sequential moves since players and enemies act in response to the choices of each other. It is also finite since either the player or the enemy will fall eventually. This sequential decision-making is based on the player's observation of patterns. Each enemy has a set of predefined attack sequences that the player can learn through repeated encounters. This is especially true in the cases of bosses whose patterns consist of multiple phases, each with distinct attacks. For example, when fighting an enemy, the player must consider the following:

- *When to attack:* The player needs to wait for an opening in the enemy's attack pattern. Striking too early may result in the enemy counterattacking, while waiting too long may allow the enemy to reset his or her position or prepare another attack.

- *When to dodge or block:* The player must time his or her defensive actions based on the enemy's movements. Dodging at the wrong time may result in taking damage while blocking too many hits can drain stamina. This can leave the player vulnerable.

Each decision in combat affects the next. For instance, if the players choose to attack and deplete their stamina, they may not have enough stamina left to dodge or block the enemy's next attack. This creates a chain of sequential decisions, where each move must be timed to maximize the player's chances of success. The outcome of each decision affects the next one and leads to a branching series of possible outcomes.

Finite-horizon sequential games can be more strongly applied to boss battles. These are regarded as some of the most iconic and challenging elements of the game. These encounters represent the ultimate test of a player's mastery of the game's mechanics and their ability to make strategic decisions under pressure. Boss fights are rarely won on the first attempt. Instead, players are expected to fail multiple times. They have to learn and adapt after each defeat. This process of iterative learning parallels evolutionary game theory, where strategies evolve over time based on the outcomes of previous encounters. An example of a boss fight is the fight against Ornstein and Smough. They are a pair of bosses that force the player to juggle two very different sets of attack patterns simultaneously. Initially, the player must decide which boss to target first. This decision hinges on the player's observations of his or her attack patterns and movement. This requires the player to anticipate how the actions of bosses will influence the next phase of the battle. However, the player does not know that defeating one boss transforms the other into a more powerful version of itself. The infamous duo is shown in **Figure 3**.

Figure 3.
Ornstein and Smough in Dark Souls—Remastered.

4.3 Iterative learning and dynamic adaptation

In evolutionary game theory, players do not necessarily have perfect information or the ability to calculate optimal strategies from the start. Instead, they learn through repeated interactions and they gradually adopt strategies that have proven effective over time. In Dark Souls, this process occurs naturally as players repeatedly face the same enemies and bosses. In the meantime, they are learning from their mistakes and adjusting their strategies accordingly. In the context of combat, but mainly in boss battles, iterative learning works as follows:

- *Initial attempt:* The player enters the fight with minimal knowledge of the enemy's behavior. Early attempts are often spent gathering information rather than attempting to win outright. These early deaths serve as learning experiences and provide the player with data on the enemy's attack patterns, range, and movement.

- *Strategy refinement:* After each death, the players refine their strategy. This might involve changing their equipment, altering their movement patterns, or choosing a different approach to managing health and stamina. Over time, the players become more efficient at recognizing and countering the enemy's attacks.

- *Learning phases:* Many bosses have multiple phases, each with distinct behaviors. For example, in the fight against Artorias the Abysswalker, the boss becomes more aggressive in the second half of the fight. This forces the player to adapt to new attack patterns. The players must learn how to handle each phase of the fight separately and have to update their strategy as the boss evolves.

Over time, the players' mental model of the boss fight becomes more complete. The information asymmetry decreases as the player becomes familiar with the boss's attack patterns. This transforms the fight from a chaotic and unpredictable encounter into a well-understood sequence of actions. Eventually, the players reach a point where they can execute their strategy with confidence. Consequently, they complete their learning process, overcome the information asymmetry, and defeat the boss.

Furthermore, the games' online community plays a role in this evolutionary process. Players often share successful strategies, item locations, and enemy weaknesses through online forums and wikis [18, 19]. This collective knowledge contributes to the evolution of the game's meta (the most effective strategies) since players adopt the strategies that have been proven effective by others. In game theory, this is similar to cultural evolution. This means that individuals learn from the experiences of others and adjust their behavior accordingly.

4.4 Dynamic difficulty and changing equilibria

Many bosses have multiple phases. Their behavior becomes more aggressive or simply change attack patterns as the fight progresses. This creates a situation where the equilibrium between the player and the boss is constantly shifting. This forces the players to adapt their strategy of bosses dynamically. For example, in the fight against Pontiff Sulyvahn, the boss summons a clone of himself to fight alongside him in the

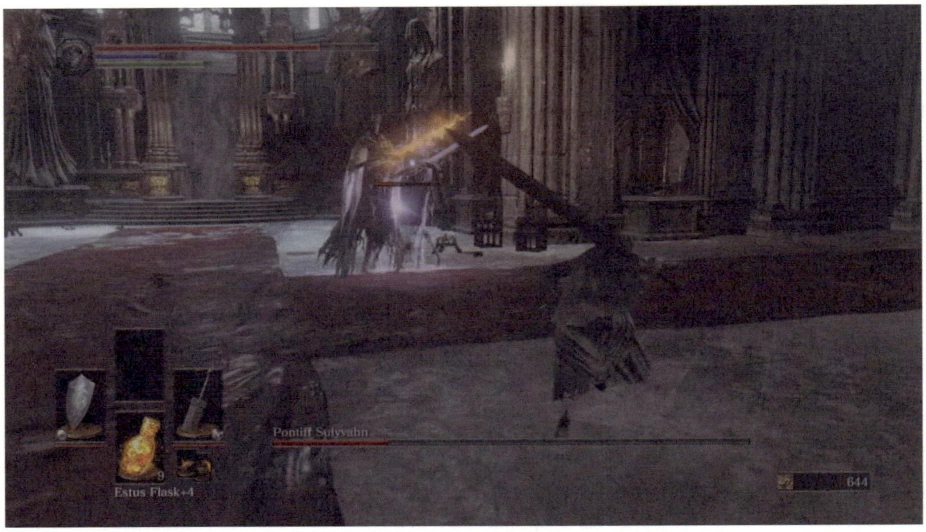

Figure 4.
The second phase of Pontiff Sulyvahn in Dark Souls III.

second phase. This change in the fight's dynamics forces the players to reassess their strategy as they must now deal with two threats simultaneously. The player's optimal strategy in the first phase of the fight (e.g., dodging and counterattacking after the boss's heavy swings) is no longer effective in the second phase. There, managing both enemies becomes the priority. This shift in the fight's dynamics can be seen as a change in the game's equilibrium. In the first phase, the player and the boss operate within a relatively stable framework and the player can rely on a consistent set of moves and responses. However, as the fight transitions to a new phase, the equilibrium shifts and the player must adapt to the new conditions. The second phase is shown in **Figure 4**.

4.5 Mixed strategies due to the unpredictability in combat

While many video games reward players for learning and exploiting enemy weaknesses, Dark Souls punishes predictability. This introduces the need for mixed strategies where players must randomize their actions to avoid becoming predictable and falling victim to enemy patterns. In game theory, mixed strategies occur when players randomize their choices to make it more difficult for opponents to predict their next move. In Dark Souls, this concept is highly relevant in PvP combat. When engaging with other human players, sticking to a repetitive strategy often leads to failure. This is due to the fact that skilled opponents will quickly recognize patterns and exploit them. Therefore, players have to adopt a mixed strategy by varying their actions (attack, dodge, block, etc.) in unpredictable ways. By keeping their strategy unpredictable, players increase their chances of success as the opponent must account for a wider range of possible actions. This type of strategy is particularly effective in zero-sum games, where one player's gain is the other player's loss. In PvP combat, every hit a player lands is a gain for him or her and a loss for the opponent. This creates a zero-sum dynamic. Both players can benefit from employing mixed strategies to outmaneuver each other.

5. Multiplayer features

While the Dark Souls games are mainly known for their single-player experience, their multiplayer features introduce an added layer of complexity and strategic depth. Multiplayer features include invasions, cooperative summoning, and PvP combat. They create a dynamic where players interact competitively or cooperatively. These interactions can be analyzed using non-cooperative game theory (where individuals compete for their own interests) and coalition game theory (where players collaborate to achieve a shared goal). Also, the multiplayer aspects of Dark Souls introduce elements of asymmetric multiplayer since players can enter each other's game worlds with varying resources, objectives, and strategies. This asymmetry presents unique challenges and opportunities for both the invaders and hosts. In doing so, the strategic decision-making that reflects the core principles of game theory is encouraged.

5.1 Non-cooperative elements in multiplayer

The invasion system allows players to invade each other's worlds to engage in PvP combat. The invader's goal is to kill the host, while the host must either defeat the invader or reach the next boss fight to force the invader out of their world. This creates a non-cooperative game where each player seeks to maximize their own payoff (survival, victory, and the acquisition of souls) at the expense of the other. This can be modeled as a zero-sum game: one player must win, while the other must lose. The asymmetry of the invasion system introduces additional strategies:

- *Resource asymmetry:* The invader often has fewer healing items than the host. This creates a disadvantage in longer encounters. This forces the invaders to adopt aggressive, hit-and-run tactics since they cannot afford to engage in extended battles where the host can heal more frequently.

- *Environmental asymmetry:* Either of the players can be familiar with the layout of the game's world and can use this knowledge to their advantage. They may lure the other into traps or areas populated by strong enemies. This can create a geographical advantage. Not to mention, enemies will attack both the invader and the host.

- *Risk-reward trade-off:* For the invaders, the risk of failure is high since they may lose their collected souls. However, the reward is also significant: by defeating the host, the invader can claim a portion of the host's souls.

In terms of strategy, the asymmetry between invaders and hosts can lead to a form of mixed strategies where the invader must rely on unpredictability to succeed. Instead of straightforward fights, invaders may adopt guerrilla tactics (surprise attacks, ambushes, and environmental hazards) to gain the upper hand. This can especially be true in areas like Blighttown or Sen's Fortress since those areas have complex terrain to trap or mislead the other player.

Another common strategy used in non-cooperative games is bluffing. Invaders can use deceptive tactics to mislead the host. For example, an invader may pretend to be low on health or act defensively to lure the host into a false sense of security. Once the host overcommits to an attack, the invader may counter with a powerful strike. Another example is disguise. Some invaders may use items like the Chameleon spell to disguise themselves as inanimate objects within the environment such as barrels or furniture. This creates uncertainty for the host who may become unsure of which object is the real invader. This is an example of a Bayesian game where players must make decisions based on incomplete information. These tactics align with the concept of strategic deception in game theory since one player manipulates the other's perception to achieve a favorable outcome. By creating ambiguity and uncertainty, the invader can force the host to make decisions under conditions of imperfect information. This, in turn, can increase the likelihood of mistakes.

5.2 Coalition in multiplayer

In contrast to invasions, the Dark Souls games also allow cooperative multiplayer interactions: players can summon others to assist them in difficult areas or boss fights. This creates an opportunity for forming temporary alliances (i.e., coalitions). Their shared goal is to defeat a boss or to clear an area. However, there are strategic considerations even within this cooperative framework:

- *Scaling difficulty:* When multiple players are summoned, the game adjusts the difficulty by increasing the health and damage output of enemies and bosses. This creates the following trade-off: summoning allies makes the fight easier in terms of strategy and coordination but harder in terms of raw numbers. The player must assess whether the added firepower and teamwork outweigh the increased difficulty.

- *Healing allies:* Some players may equip spells or items that allow them to heal their allies. In turn, they sacrifice their own resources to keep the team alive. This creates a trade-off between personal risk (losing valuable resources) and the collective benefit of keeping the team in the fight.

- *Dividing roles:* In many cooperative boss fights, players will take on specific roles such as one player drawing the boss's attention (tanking) while others focus on dealing damage or healing others. This division of labor mirrors the concept of role specialization in cooperative game theory where players adopt complementary strategies to maximize the overall success of the group.

- *Division of rewards:* In cooperative boss fights, all summoned players receive a portion of the souls and rewards earned from defeating the boss. However, the host of the world receives the majority of the rewards, while the summoned players receive a smaller share. This distribution of rewards mirrors coalition game theory where the central player (the host) benefits more from the coalition than the peripheral players (the summoned allies).

5.3 Invasions and coalitions at the same time

A variation on the invasion mechanic occurs when an invader enters the world of a host who has previously summoned allies. This creates a multi-agent dynamic: the host and his or her summoned allies must work together to defeat the invader. Regarding game theory, this situation introduces elements of multi-agent cooperation within a competitive framework.

In this case, the invader faces a large disadvantage due to being outnumbered. It may force the invader to adopt a divide-and-conquer strategy. The invader's goal is to separate the host from his or her allies. This, in turn, reduces the coalition's effectiveness and increases the invader's chances of defeating each player individually. This mirrors coalition-breaking strategies in game theory where a lone player attempts to disrupt an established alliance to shift the balance of power. For the host and his or her allies, success lies in maintaining their coalition and coordinating their actions to prevent the invader from exploiting any weaknesses. This requires clear communication and a shared understanding of the game. The challenge of maintaining cooperation under pressure is similar to the principles of cooperative game theory since the stability of a coalition is tested by external threats.

6. Information asymmetry, uncertainty, and meta-learning

In many video games, success is based on how much information the players have at their disposal. Players are often given maps, clear objectives, and detailed enemy information. However, in the Dark Souls games, these are not available for the players. Therefore, they are provided limited information and they are left to uncover the world and its dangers mostly on their own. This lack of explicit guidance and the constant presence of information asymmetry force players to make decisions under uncertainty. This is a concept that is very similar to Bayesian game theory and decision theory.

6.1 Information asymmetry

In game theory, information asymmetry occurs when one player has more or better information than another player or when the players do not have full knowledge of the game environment. In Dark Souls, information asymmetry plays a critical role in shaping the players' experience. The game withholds important information from the players. In doing so, they are forced to learn through exploration, experimentation, and repeated failure. The following aspects of these games contribute to information asymmetry:

- *Hidden enemies and traps:* Throughout the game, enemies and traps are often hidden in corners, sometimes behind doors, or lurking around unseeable parts of the area, just waiting to ambush the player. For example, the first game's infamous Anor Londo archers are a deadly threat to players who first encounter them. They are strategically placed to surprise and overwhelm the players. There are environmental hazards as well since the players can easily fall to their deaths. These further complicate the players' decision-making process. This part of the game is shown in **Figure 5**.

- *Oblique lore and storytelling:* Unlike many role-playing games, the Dark Souls games do not rely heavily on cutscenes or explicit storytelling. Much of the games' lore is hidden in item descriptions, environmental cues, and cryptic dialog. Players must piece together the narrative themselves. This creates a sense of discovery but also a profound sense of uncertainty about their purpose and goals.

- *Unmarked shortcuts and secret areas:* The world of Dark Souls is interconnected. There are numerous shortcuts and secret areas that are not immediately obvious to the player. These hidden paths often lead to significant rewards, but they can also lead to additional dangers.

Figure 5.
The infamous archers in Anor Londo (Dark Souls). Two archers are in front of and firing at the player.

This asymmetry creates an environment where the players are constantly operating with incomplete information. Their decisions must be made without full knowledge of what lies ahead, whether it be a powerful enemy, a hidden trap, or a valuable item. As mentioned, these scenarios align with Bayesian decision-making in game theory since players must make choices based on the limited information they have. In the meantime, they have to update their beliefs and strategies as they gather more data through experience. This is also similar to the iterative learning method as mentioned in Section 4.3.

For example, consider the decision to enter a new area of the game. The players do not know what enemies or hazards await them in this area. However, they may have a prior expectation based on the layout and difficulty of the previous areas. If the area looks treacherous, they might proceed with caution. Perhaps, they may even choose to use ranged attacks or just scout ahead cautiously. As they explore the area and encounter enemies or traps, they gather more information and they update their strategy. Using the Bayesian updating process, the players start with a hypothesis (e.g., this new area might contain strong enemies or traps) and adjust that hypothesis based on the outcomes they experience (e.g., encountering or defeating a specific enemy). Over time, the players build a mental model of the game world while learning to recognize environmental cues that signal danger or rewards. For instance, after encountering a hidden enemy ambush in one area of the game, the players may learn to approach similar areas more cautiously in the future. This is a form of Bayesian learning since the players update their strategy based on their past experiences. They are gradually improving their ability to predict and react to the game's challenges.

6.2 Fear of the unknown

Another aspect of information asymmetry in Dark Souls is the fear of the unknown. The game's world is designed to evoke a sense of unease and tension. This means that every corner may hide a new danger and the player is never sure what to expect next. This uncertainty creates a psychological dynamic where the player must constantly assess the risk of exploring new areas. In terms of game theory, this can be viewed as a decision under uncertainty where the player must choose between two or more actions with unknown outcomes. For instance, the player might encounter a fork in the road: one path leads into a dark cave, and the other leads to a more open, well-lit area. The player has limited information about what lies down either path and must decide which way to go based on the potential risks and rewards.

The tension between exploration and safety is central to the players' experience in the Dark Souls games. The desire to uncover hidden treasures or shortcuts must be balanced against the risk of encountering enemies or traps. This trade-off creates a payoff matrix in game theory where each decision to explore carries both potential rewards (discovery of valuable items or new areas) and risks (losing souls or health). The players' risk tolerance will shape their approach to exploration: some players might be more cautious and calculate risks, while others may choose to explore more aggressively.

6.3 Meta-learning

Besides each player's individual journey and personal learning process, there is also a meta-level of learning that occurs through the games' online community. One of the reasons why Dark Souls has such a strong and engaged community is its unforgiving difficulty. Players often share information about hidden items, boss strategies, and optimal builds through online forums, wikis, and YouTube tutorials. This collective knowledge helps to mitigate the information asymmetry experienced by individual players and allows them to benefit from the experiences of others. In game theory, this is similar to common knowledge where information that is shared among players affects how they make decisions. As players discover new strategies or uncover hidden aspects of the game, this knowledge becomes part of the shared understanding of the community. This, in turn, reduces the overall uncertainty for future players. For example, a player who is struggling with the boss Ornstein and Smough might consult a community guide that suggests defeating Ornstein first to reduce the difficulty of the second phase. By using this information, the player is able to reduce the information asymmetry and approach the fight with a more effective strategy. In this way, the collective knowledge of the community helps to level the playing field and allows the players to make more informed decisions and overcome the challenges of the game more efficiently.

In addition to individual learning and iterative adaptation, the development of meta-strategies in Dark Souls can be understood as a form of cultural evolution. This means that successful strategies are passed down from experienced players to newcomers. Similarly to how evolutionary game theory explains the spread of successful behaviors in populations, cultural evolution in Dark Souls describes how certain strategies become dominant in the player community over time. For example, certain builds (combinations of character stats, equipment, and abilities) may become more popular within the community as players discover their effectiveness. For instance, a build that focuses on high stamina and dexterity allows for fast attacks and dodges. This might become the dominant strategy for defeating bosses that have slow, telegraphed attacks. As more players adopt this build and share their success stories, the strategy spreads through the community. This creates a meta, which is the set of strategies considered most effective by the player base at any given time.

This process of cultural transmission is accelerated by the interconnected nature of the online Dark Souls community. Players not only use strategies from others but also experiment with their own variations since they might refine and optimize them through their own experiences. Over time, the community collectively evolves its understanding of the game and each new discovery contributes to the refinement of the meta. In evolutionary game theory, this is similar to the process of strategy refinement where players continuously adjust their tactics based on the success or failure of past decisions.

However, it should be noted that new patches or updates might be released by the developers over time. Some patches may contain balance adjustments; therefore, players may need to adapt their strategies once again. This creates a dynamic and evolving meta. For instance, a weapon or build may become overpowered due to a game update. This may lead to many players adopting it as the dominant strategy. However, in response to community feedback, the developers may release a patch that nerfs the weapon's damage or effectiveness. This can also force the players to seek alternative strategies. This dynamic between developer-driven changes and community adaptation creates a feedback loop where the meta is constantly evolving.

6.4 The influence of shared knowledge

The influence of shared knowledge on meta-strategies is evident in the way players approach boss fights and difficult areas. As players share information about optimal strategies, these approaches become widely adopted. For example, certain cheese strategies (tactics that exploit the game's mechanics to defeat enemies or bosses with minimal effort) often spread through the community. While not always intended by the developers, these strategies offer players an easy way to bypass particularly difficult sections of the game. One well-known example is the Iron Golem boss fight in the first Dark Souls game. There, players can manipulate the movement of the boss to make it fall off a ledge. This can end the fight almost immediately. Once this strategy was discovered, it quickly became part of the community's shared knowledge. Therefore, many players started using it as an easy way to defeat the boss.

These shared strategies affect how the game is played at a meta-level as the community begins to favor certain tactics or builds that provide the highest chance of success. This phenomenon is similar to the way strategies evolve in competitive games where players adopt and refine the most effective approaches based on the collective knowledge of the community. However, the development of a meta can also reduce the diversity of strategies used by players. As certain strategies become dominant, others may start to fade from the collective knowledge. This can lead to a more homogeneous style of play. In terms of game theory, this represents a shift toward an equilibrium strategy since players converge on a specific set of tactics that are considered optimal for the given game environment. While this can lead to more efficient play, it may also reduce the variety of experiences that players have. In the meantime, the community converges toward the same set of solutions.

7. Conclusions

In this study, we have explored how various aspects of game theory can be applied to the Dark Souls games. The Dark Souls series creates a game world where players are forced to think strategically, adapt to new information, and optimize their decision-making processes. The series' combination of challenging mechanics, high stakes, and player behavior provides a case study for understanding how players engage with difficult scenarios and develop effective strategies. By examining these games in the context of game theory, we can gain a deeper understanding of the following dynamics:

- *Sequential decision-making:* Combat and boss fights in Dark Souls are inherently finite-horizon sequential games. Players have to anticipate enemy behavior, manage their resources, and react to evolving situations in real time. This is essential for understanding the rhythm of gameplay since each action has consequences that affect subsequent decisions.

- *Iterative learning and adaptation:* Players have to continuously refine their strategies after each failure. Through repeated encounters with enemies and bosses, players gather new information, adjust their tactics, and improve over time. This aligns with evolutionary game theory in which strategies evolve based on success or failure.

- *Risk management and payoff structures:* Every decision in Dark Souls has risk and reward mechanics. The payoff structures of the series force players to consider both short-term and long-term consequences. They have to weigh the risks of their actions against potential rewards.

- *Information asymmetry and uncertainty:* The emphasis on exploration and discovery creates information asymmetry since players are often unaware of what lies ahead of them. This uncertainty forces them to make decisions with incomplete information, which is similar to Bayesian decision-making in game theory. Over time, players learn to reduce this asymmetry through observation, experience, and the use of external resources like guides and community knowledge.

- *Meta-learning and shared knowledge:* The player community shares information and strategies online. This collective knowledge helps players overcome challenges more efficiently and highlights the importance of external resources in reducing difficulty. This reflects how player behavior evolves in response to both game mechanics and community.

Beyond its applications for understanding how players engage with existing games, game theory can offer an understanding for game designers. By incorporating elements of game theory into their video games, developers can create more balanced, engaging, and challenging game experiences. Some potential applications of game theory in designing video games include balancing difficulty and player choices, designing competitive and cooperative systems, and encouraging iterative learning.

In conclusion, the principles of game theory help explain why the Dark Souls games continue to captivate players and create a community of players. By forcing players to make difficult choices, adapt to changing conditions, and engage with uncertainty, the Dark Souls games offer an experience that challenges players not only physically but also mentally. These same principles can be applied to other games and strategic scenarios, which shows the relevance of game theory in both digital and real-world contexts.

Acknowledgements

This work has been implemented by the TKP2021-NVA-10 project with the support provided by the Ministry of Culture and Innovation of Hungary from the National Research, Development and Innovation Fund, financed under the 2021 Thematic Excellence Programme funding scheme.

Conflict of interest

The authors declare no conflict of interest.

Author details

Tibor Guzsvinecz* and Judit Szűcs
Department of Information Technology and Its Applications, Faculty of Information Technology, University of Pannonia, Zalaegerszeg, Hungary

*Address all correspondence to: guzsvinecz.tibor@zek.uni-pannon.hu

References

[1] Guzsvinecz T. The Soulsification of video games. Multimedia Tools and Applications. 2024:1-27. DOI: 10.1007/s11042-024-19628-4

[2] Anderson CG, Dalsen J, Kumar V, Berland M, Steinkuehler C. Failing up: How failure in a game environment promotes learning through discourse. Thinking Skills and Creativity. 2018;**30**:135-144. DOI: 10.1016/j.tsc.2018.03.002

[3] Horsfall M, Oikonomou A. A study of how different game play aspects can affect the popularity of role-playing video games. In: 2011 16th International Conference on Computer Games (CGAMES). Louisville, KY, USA: IEEE; 2011. pp. 63-69. DOI: 10.1109/CGAMES.2011.6000361

[4] Gandolfi E. Enjoying death among gamers, viewers, and users: A network visualization of Dark Souls 3's trends on Twitch.tv and Steam platforms. Information Visualization. 2018;**17**(3):218-238. DOI: 10.1177/1473871617717075

[5] Pan S, Xu GJW, Guo K, Park SH, Ding H. Cultural insights in souls-like games: Analyzing player behaviors, perspectives, and emotions across a multicultural context. IEEE Transactions on Games. 2024:1-12. DOI: 10.1109/tg.2024.3366239

[6] Ford D. Giantness and excess in dark souls. In: International Conference on the Foundations of Digital Games. New York, NY, USA: ACM; 2020. pp. 1-4. DOI: 10.1145/3402942.3402966

[7] Broom M, Rychtář J. Game-Theoretical Models in Biology. Boca Raton: Chapman and Hall/CRC; 2022. DOI: 10.1201/9781003024682

[8] Krueger JI, Heck PR, Evans AM, DiDonato TE. Social game theory: Preferences, perceptions, and choices. European Review of Social Psychology. 2020;**31**(1):222-253. DOI: 10.1080/10463283.2020.1778249

[9] Guardiola E, Natkin S. Game theory and video game, a new approach of game theory to analyze and conceive game systems. In: Proceedings of the Seventh International Conference on Computer Games: AI, Animation, Mobile, Educational & Serious Games. Paris, France: Center for Studies and Research in Computer Science and Communication (CEDRIC); 2005. pp. 166-170

[10] Theodorou A. Prepare to die: Reconceptualising death, and the role of narrative engagement in the Dark Souls series (2011-2018). In: Death, Culture & Leisure: Playing Dead. Leeds, UK: Emerald Publishing Limited; 2020. pp. 83-94. DOI:10.1108/978-1-83909-037-020201013

[11] Neto A, Cardoso P, Carvalhais M. Asynchronous interactions between players and game world. In: Springer Series in Design and Innovation. Cham: Springer International Publishing; 2021. pp. 148-156. DOI: 10.1007/978-3-030-61671-7_14

[12] Andiloro A. Understanding genre as atmospheric assemblage: The case of videogames. Television and New Media. 2023;**24**(5):559-570. DOI: 10.1177/15274764231171076

[13] Guzsvinecz T. The correlation between positive reviews, playtime, design and game mechanics in souls-like role-playing video games. Multimedia Tools and Applications.

2023;**82**(3):4641-4670. DOI: 10.1007/s11042-022-12308-1

[14] Kunzelman C. How we deal with dark souls: The aesthetic category as a method. In: Hybrid Play. Abingdon, Oxon, UK: Routledge; 2020. pp. 167-181

[15] Vorderer P, Klimmt C, Ritterfeld U. Enjoyment: At the heart of media entertainment. Communication Theory. 2004;**14**(4):388-408. DOI: 10.1111/j.1468-2885.2004.tb00321.x

[16] Kahneman D, Tversky A. Prospect theory: An analysis of decision under risk. In: World Scientific Handbook in Financial Economics Series. Singapore: World Scientific; 2013. pp. 99-127. DOI: 10.1142/9789814417358_0006

[17] Kalmanlehto J. Aesthetics of agency and the rhythm of gameplay. Games and Culture. 2024;**19**(8):1038-1054. DOI: 10.1177/15554120231185630

[18] Robinson B, Czauderna A, von Gillern S. "I think I get why y'all do this now": Reckoning with Elden Ring's difficulty in an online affinity space. Games and Culture. 2023. DOI: 10.1177/15554120231203134

[19] Czauderna A, von Gillern S, Robinson B. Informal social-emotional learning in gaming affinity spaces: Evidence from a Reddit discussion thread on Elden Ring. Simulation and Gaming. 2024;**55**(1):30-50. DOI: 10.1177/10468781231209697

Section 3

Game Theory in Resource Management and Allocation

Chapter 5

Game Theory in Water Quality Management

Ali Moridi

Abstract

Game theory systematically examines competition and cooperation among stakeholders with conflicting objectives, such as managing downstream water supply and waste load allocation. Games are defined as mathematical models comprising players, available strategies, and payoffs for each strategy combination. These payoffs determine decisions and game types as non-zero- and zero-sum games. Game theory-based issues are mostly multi-decision-maker and multi-criteria problems. Most problems are simplified in a conventional optimization model with a single decision-maker and a unified objective model. This approach assumes perfect cooperation among decision-makers, who optimize the overall objective without prioritizing individual goals. However, game theory maintains that each decision-maker optimizes their own objective, considering the mutual influence of decisions. Game theory predicts stable, but not necessarily Pareto-optimal outcomes, as players focus on maximizing their benefits. It realistically simulates stakeholder behavior, often resulting in non-cooperative behavior even when there is more beneficial cooperation. In water quality conflict resolution, game theory models interest groups as decision-makers, whose combined choices determine conflict outcomes. It provides a framework for analyzing water conflicts and suggests innovative solutions for better resolution.

Keywords: water resource management, Nash bargaining theory, bankruptcy, dilemma of prisoners, socialist cooperative game, chicken game, stag hunt game

1. Introduction

The quality of water resources, including seas, lakes, reservoirs, rivers, and groundwater, significantly impacts water usage. This is especially crucial in regions where water scarcity and supply limitations are prevalent. The main part of water quality management is controlling human activity pollution sources, especially in areas where pollutants have severely degraded water quality. Pollution control levels are determined based on water quality standards of different water consumption.

Water pollution occurs when wastewater discharges disrupt the natural ecological balance of a water body. Point sources of pollution are discharged into the sea, lake, surface, and groundwater from a single point (like industrial and domestic wastewater). Domestic sewage, which comes from homes, commercial establishments, and public institutions, is one type of point source. Conversely, non-point sources release pollutants from multiple locations, including agricultural return flows and urban runoff.

While reducing or eliminating point source pollution can be achieved through proper treatment processes before discharge, treating non-point source pollution is often not economically feasible. However, numerous structural and non-structural Best Management Practices (BMPs) can reduce pollution from different non-point sources.

One common approach to water quality management involves applying various optimization techniques, particularly linear programming, to suitable objective functions. These optimization methods typically assume that minimizing costs is the primary motivation for all pollution sources. Environmental parameters can be included in these objective functions by converting them into monetary equivalents. Since each agent aims to reduce its own pollution control costs, often at the expense of others, this approach positions all water pollution sources within a competitive framework. Non-cooperative game theory is particularly well-suited for modeling such frameworks, where groups of agents focus on minimizing their individual pollution reduction costs. However, cooperation, rather than pure competition, can sometimes play a more significant role in water quality management, as most community members share common environmental concerns.

2. Total maximum daily load (TMDL)

TMDL is the basis for incorporating game theory into water quality management. Initially, the primary goal of river pollution control was to protect public health, particularly in areas where large cities and industries are located near rivers. However, in recent years, the focus has expanded to include the protection of ecosystems and water resources and preserve the natural environment associated with these resources. Consequently, the main objective of water quality management is to control pollutants discharged into water bodies to ensure that river water quality at critical points does not fall to unacceptable levels below the natural background level.

TMDL is the maximum pollution load (point and non-point pollution sources) that can be discharged to a waterbody while meeting those pollutant water quality standards. It establishes a target for pollution reduction, and then, pollution load reduction of each pollutant source can be determined by game theory methods.

Additionally, the TMDL specifies the maximum allowable pollution load that can be discharged into the waterbody. Pollutant sources are classified as non-point sources and point sources, which receive a load allocation (LA) and waste load allocation (WLA), respectively. The water quality seasonal variations are considered in TMDL. So, in order to decrease violations of water quality standards and in order to account for uncertainties in predicting the effectiveness of pollutant reductions, a margin of safety (MOS) is considered.

Mathematically, the TMDL equation is expressed as:

$$TMDL = \sum ULA + \sum NPLA + \sum WLA + MOS \tag{1}$$

Where ULA is the sum of load allocations to upstream pollution and background, NPLA is the non-point source pollution load allocation, WLA is the point source waste load allocation, and MOS is the safety margin.

Any pollutant, whether from a point or non-point source, which impairs or threatens a waterbody is referred to as a waterbody-specific pollutant combination. Usually, a TMDL is established for each combination. After determining the TMDL, allocating pollution reduction responsibilities to point and non-point sources becomes necessary.

This allocation process is often contentious and can be analyzed using game theory. The apportioned pollution load directly influences wastewater treatment costs for point sources and the costs of pollution reduction of non-point sources by implementing BMPs [1].

3. Classical water quality management: Waste load allocation optimization model

In this section, water quality management and optimization models based on pollution reduction and cost minimization are presented. The objective function of the proposed model is cost minimization of pollution control (wastewater treatment for point source and BMP for non-point source) of all point and non-point water pollution sources. The optimization model structure is as follows [2–4]:

$$\min Total\ Cost = \sum_{i=1}^{n} cost_i\left(pr_i\right) \tag{2}$$

Subject to:

$$\sum_{i=1}^{n} x_{i,j} * \left(1 - pr_i\right) \le Standard_j \qquad j = 1,2,3,\ldots,m \tag{3}$$

Where:

$$LB_i \le pr_i \le UB_i \tag{4}$$

$$x_{i,j} = function\left(pr_i,\ adv\ and\ dif\ of\ pollution\ in\ water\ bodies\right) \tag{5}$$

pr_i: Percent of pollution reduction of pollution source i.

$cost_i\, pr_i$: Cost of pollution reduction (wastewater treatment for point source and BMP for non-point source) for pollution source i.

n: Total number of point and non-point pollution sources.

xij: Transfer coefficient between pollutant source i and water quality monitoring node j, which is calculated based on water quality modeling and indicates the impact of pollution reduction at source i on water quality node j.

m: Total number of control points based on the water quality monitoring network.

$Standard_j$: Water quality standard at control point j. Based on the violation from TMDL, the required water pollution reduction at control node j will be considered in the optimization model.

LB_i/UB_i: Lower bound/upper bound of pollution reduction of point and non-point pollution source i based on wastewater treatment technology and BMPs.

4. Water quality management in lakes and reservoirs

The advection and diffusion of water quality variables in lakes and reservoirs differ significantly from those in rivers and streams. Reservoirs can dramatically alter the characteristics (biological, chemical, and physical) of incoming water. Nutrients, particularly phosphorus, are the primary pollutants responsible for eutrophication, which can severely degrade lake water quality. Biological and chemical oxygen demand from pollution discharges is also crucial, especially when the pollutants originate from domestic or industrial sources entering a lake or upstream river [5, 6].

Density differences caused by water temperature, dissolved substances, and suspended solids can lead to the stratification of water in lakes and reservoirs. This vertical stratification affects the biological and chemical characteristics of a lake or reservoir and influences the advection and diffusion of water quality variables, resulting in varying water quality at different depths. The thermocline hinders the diffusion of dissolved oxygen and other substances, isolating the hypolimnion (bottom layer) from the epilimnion (upper layer) and the atmosphere, potentially leading to oxygen depletion. This condition can result in the production and release of methane and hydrogen sulfide from the lake or reservoir bottom, impacting water quality at different layers and allowing for selective water withdrawal to meet downstream quantity and quality demands, while also severely degrading the aquatic environment.

Eutrophication is another major concern in lakes and reservoirs, disrupting their ecological balance. Nutrient input increases productivity and nutrient richness, leading to eutrophic conditions characterized by high nutrient levels, abundant algae, and reduced water clarity. Nitrogen and phosphorus are the primary drivers of algal growth and eutrophication. Consequently, water quality management focuses on limiting nutrient input to slow down eutrophication, primarily by controlling phosphorus levels. Reservoir water quality often deteriorates over time due to thermal stratification and pollutant accumulation, impacting downstream water quality. Selective withdrawal structures are effective in managing the quality of these releases [7, 8].

Game theory can be applied to two kinds of conflicting objectives in reservoir and lake water quality management. The first is related to river-reservoir water quality management, which involves reducing the upstream river pollution load to increase water quality index in lakes and reservoirs. In these conflicts, the maximum pollution load discharged into the lake must be evaluated, and the upstream total maximum daily load (TMDL) must be calculated accordingly. Subsequently, waste load allocations can be defined using various methods, particularly game theory approaches, which will be discussed later.

The second conflicting objective involves water allocation to downstream demands with desirable quality and standards. In this scenario, conflicts between different downstream water demands and the reservoir water quality are the main issues which will be discussed later [9, 10].

5. A systematic approach to resolving conflicts in water quality management

Water quantity and quality management is a complex endeavor that necessitates balancing water distribution with acceptable standards while addressing conflicts among diverse stakeholders. These stakeholders encompass both those directly affected by decisions, such as water users and polluters, and those with the power to influence outcomes, including government agencies and environmental organizations. Key factors contributing to water conflicts include imbalances in water quantity and quality, competing water uses, the involvement of various governance bodies, and the stage of national development [11].

Water quality management frequently clashes with urban and rural development, agriculture, industry, and other economic sectors. The allocation of water for domestic, industrial, and agricultural purposes often conflicts with preserving essential water flows to support aquatic ecosystems. The quality of wastewater discharged from domestic, industrial, and agricultural sources significantly impacts overall water quality, influenced by the natural purification capacity of water bodies. A perpetual tension exists between polluters seeking to minimize treatment costs by increasing waste load allocations and downstream users and environmental regulators demanding higher water quality standards [12].

Traditional conflict resolution methods, such as legal and administrative processes, often favor one party over others. Contemporary water management increasingly employs game theory to incorporate the perspectives of all stakeholders in resolving disputes. This cooperative approach involves four stages:

Problem definition: Clearly define the issue, identify key stakeholders, and understand their goals and concerns [13, 14].

Systems mapping: Create a shared understanding of the conflict by visualizing relationships between stakeholders and issues.

Cooperative dialog: Facilitate open communication among stakeholders to identify common ground and address underlying tensions.

Action planning: Develop and implement collaborative solutions based on shared objectives.

A systems perspective is essential for effective water management. It involves:
Understanding relationships: Analyzing the interconnectedness of physical, ecological, and social components.

Characterizing components: Describing the attributes of system elements, including human values and behaviors.

Assessing impacts: Evaluating the consequences of actions on both the environment and society.By adopting a systems approach and employing collaborative conflict resolution strategies, it is possible to achieve more sustainable and equitable water management outcomes.

6. Game theory

Water is a vital resource that requires efficient management, particularly in contexts involving multiple stakeholders with competing interests. These stakeholders often include governments, industries, agricultural sectors, and local communities. Game theory, a mathematical framework for analyzing strategic interactions between players, offers a powerful tool for resolving conflicts and promoting cooperation in water resource management. In water resources, game theory can help identify optimal strategies for allocating and sharing water among various users while minimizing conflicts. This chapter will explore the application of both zero-sum and non-zero-sum games in water resource management, including relevant examples to illustrate their practical uses. In game theory, the key components are players (the stakeholders or decision-makers) and strategies (the available options or actions they can take). Each player seeks to maximize their payoff, which depends not only on their own choices but also on the choices of others [15, 16].

6.1 Zero-sum and non-zero-sum games

A zero-sum game is one in which the gain of one player directly corresponds to the loss of another. The total "payoff" is constant, and no cooperation between players is possible. These games often represent highly competitive or adversarial situations. In zero-sum water resource management scenarios, two or more stakeholders compete over a fixed quantity of water, and one party's gain is equal to another's loss. This can occur in situations where there is little or no room for cooperation and the goal is purely to secure water at the expense of others.

One classic example of a zero-sum game in water resources is the upstream-downstream conflict between two regions sharing a river. The upstream region may wish to divert water for agricultural purposes, while the downstream region requires water for municipal or industrial uses.

Players: Upstream and downstream stakeholders.

Strategies: The upstream region can either divert more or less water for irrigation, while the downstream region can invest in water-saving technologies or rely on natural river flow.

Payoff: If the upstream region diverts more water, it benefits at the expense of the downstream region, and vice versa.

In this scenario, cooperation is difficult because both players view the resource as a fixed, scarce commodity, leading to a competitive dynamic. While zero-sum games offer insights into highly competitive scenarios, they may oversimplify the complexity of water resources management, where cooperation is often possible and desirable.

Zero-sum thinking can exacerbate conflicts and reduce opportunities for mutually beneficial solutions.

Non-zero-sum games occur when the interests of players are not strictly opposed, allowing for the possibility of cooperation. In such games, both players can gain or lose simultaneously, and cooperative strategies can lead to better overall outcomes. Non-zero-sum games are more common in water resources management. In these games, the players' actions can lead to outcomes where all parties either gain or lose together. The possibility of cooperation means that strategies can be developed to optimize the use of water for all stakeholders. Consider a river shared by two countries that both depend on it for agriculture, industry, and domestic water use. In this non-zero-sum scenario, cooperation can enhance water availability for both countries.

Players: Country A and Country B.

Strategies: Each country can either over-extract water for short-term benefits or cooperate on sustainable water management strategies, such as building joint infrastructure (e.g., dams or irrigation systems) or implementing water-sharing agreements.

Payoff: If both countries cooperate, they can achieve a sustainable flow of water that benefits agriculture, industry, and domestic needs for both. If one or both countries choose non-cooperation, they may face long-term water shortages or environmental degradation, which reduces overall benefits [17].

One practical application of non-zero-sum game theory is in the negotiation of water-sharing treaties between countries that share transboundary rivers. These treaties often involve complex negotiations where both sides seek to maximize their water use while avoiding conflict. Game theory can be used to model these negotiations, identify equilibrium strategies, and facilitate cooperative agreements that balance the needs of all parties. The Nile River Basin provides an excellent example of a non-zero-sum game in water management. Countries along the Nile, such as Egypt, Sudan, and Ethiopia, have historically had conflicting interests in the use of the river's waters. Ethiopia's construction of the Grand Ethiopian Renaissance Dam (GERD) has introduced new challenges to water sharing. While Egypt is concerned about reduced water availability, Ethiopia aims to generate hydroelectric power for economic development. Game theory helps model the situation as a non-zero-sum game, where cooperative strategies—such as negotiating water release schedules—can benefit both countries.

Players: Egypt, Sudan, Ethiopia.

Strategies: Negotiating dam operations, water-sharing agreements, or unilateral actions.

Payoff: Cooperation could lead to economic growth in Ethiopia while ensuring water security for Egypt and Sudan. Non-cooperation could lead to regional tensions and reduced water security for all [18].

6.2 Cooperative and non-cooperative games

Many real-world water resource conflicts exhibit characteristics of both zero-sum and non-zero-sum games. For example, in the short term, competing users may view water as a fixed resource, leading to zero-sum competition. However, in the long term, investments in infrastructure, technology, and cooperative governance may turn the situation into a non-zero-sum game. An example is the allocation of water between agricultural and urban sectors during droughts. In times of water scarcity, the situation may seem zero-sum as both sectors compete for a fixed resource. However, with cooperative management strategies such as water recycling or better irrigation techniques, both sectors can benefit, making it a non-zero-sum game over time. Game theory offers a structured way to analyze the strategic interactions involved in water resource management, providing insights into both competitive and cooperative dynamics [19]. Zero-sum games highlight conflicts where one party's gain is another's loss, while non-zero-sum games reveal the potential for mutual gains through cooperation. In practice, water resource conflicts often involve a combination of both game types. By understanding the incentives and strategies of all players, policy-makers and water managers can use game theory to design better water allocation systems, foster cooperation, and avoid destructive conflicts [20]. So, we can say:

- Zero-sum games are suitable for analyzing competitive, conflict-driven scenarios where water is seen as a fixed commodity.
- Non-zero-sum games are ideal for modeling cooperative strategies that lead to sustainable water management.
- The combination of zero-sum and non-zero-sum elements often better reflects the complexity of real-world water management issues.

In non-cooperative games, stability can be understood through several frameworks, including Nash equilibrium, Sequential Stability (SEQ), General Meta-Rationality (GMR), and Symmetric Meta-Rationality (SMR). These stability types can be classified using three key factors:

- Foresight, which refers to the number of steps a player considers when making decisions.

- Willingness to accept disimprovement, where a player might choose a strategy that reduces their own payoff to negatively impact others.

- The player's awareness of other participants' priorities [21].

The Nash equilibrium is the most commonly applied concept for identifying stable strategies in non-cooperative games. It assumes players are fully rational and lack foresight, with each player selecting a strategy that maximizes their own payoff. A Nash equilibrium is reached when no player can improve their outcome by changing their strategy. If the strategy choice j is the Nash response for every player, then it becomes the solution for the entire game. However, a major limitation of the Nash equilibrium is that it does not account for players' awareness of their opponents' strategies, making it difficult to predict precise outcomes. As a result, additional theories have been introduced to build upon Nash's concept and incorporate opponents' reactions into decision-making [22].

General Meta-Rationality (GMR) offers another stability concept in non-cooperative games. A state is considered GMR-stable if any unilateral action that improves a player's payoff is constrained by opponents' responses. Symmetric Meta-Rationality (SMR), a more restrictive form of GMR, focuses on situations where a player's strategy change from x to z is constrained by the opponent, and no alternative strategy y yields a higher payoff than x for the player. Sequential Stability (SEQ) resembles GMR but with a key distinction: Players only make moves that improve their own payoffs without intending to reduce their opponents' outcomes, reflecting both conservatism and risk-taking behavior [23].

7. Nash bargaining theory

As previously discussed, the involvement of multiple decision-makers considerably complicates the decision-making process due to divergent priorities and objectives. Reconciling these conflicting goals often presents a formidable challenge.

Several strategies can be employed to address conflict situations. One approach involves framing the issue as a multi-objective optimization task that includes the goals of various decision-makers. Alternatively, conflict situations can be modeled as social choice problems, taking into account the preferences of all stakeholders in decision-making. Nash proposed a new approach, defining specific conditions for a solution and demonstrating that a singular solution meets these fairness criteria. This part delves into the Nash Theory [24].

Consider a conflicting water quality problem with n stakeholder/decision-makers.

Let $x_i \in S$ represent the decision variables, and $f_i(x_i)$ denote the utility function of stakeholder i. The criteria decision space is defined as:

$$U = \{u_i | u_i \in U^n, u_i = f_i(x_i), x_i \in S\} \tag{6}$$

Assuming no agreement is reached, all stakeholders will receive low utility values.

Let d_i represent disagreement point for stakeholder i. Consequently, the conflict is characterized by the pair of U and d, where U represents the set of all potential outcomes and d signifies the disagreement point. Any solution to the conflict is thus a function of U and d, denoted as (U,d). This solution function is subject to the following situation:

Feasibility: The solution must be attainable within the feasible set: (U, d) $\in$ U.

Individual rationality: The solution must yield the disagreement point for all stakeholders: (U, d) $\geq d$.

Pareto optimality: The solution must be non-dominated, meaning no other feasible. Solution can improve the outcome for one stakeholder without harming another.

Independence of irrelevant alternatives: A feasible solution should remain unchanged if some options become infeasible.

Scale invariance: Linear transformations of the objective functions should not alter the solution.

Symmetry: If two stakeholders have identical locations in the conflict, they should receive the same outcomes in the solution.

Under the conditions of U being convex, closed, bounded, and containing a dominant element over d ($\underline{u} > \underline{d}$), a unique solution ($f^*$ (U, d)) is guaranteed and can be determined by solving the following optimization problem:

$$\text{Ṭh} \quad \sum_{i=1}^{n} (u_i - d_i) \tag{7}$$

Subject to:

$$u_i = f_i(x_i), \; x_i \in S \tag{8}$$

$$u_i \geq d_i \tag{9}$$

$$u_i \in U^n \tag{10}$$

The objective function is termed the Nash product. Importantly, this approach can be interpreted as a distance-based method that maximizes geometric distance from the disagreement point d.

In practical scenarios, stakeholders often possess varying degrees of power. To account for these differences, Eq. (7) can be modified as:

$$\text{Ṭh} \quad \sum_{i=1}^{n} (u_i - d_i)^{w_i} \tag{11}$$

Here, w_i represents the relative power of decision-maker i. The non-symmetric Nash bargaining solution is the answer to this problem. **Figure 1** illustrates typical utility functions for allocated pollution discharge and water quality at control points of different stakeholders, ranging from zero to one within the allocated water range of b to c or reducing from a to b by increasing the pollution at control points [25].

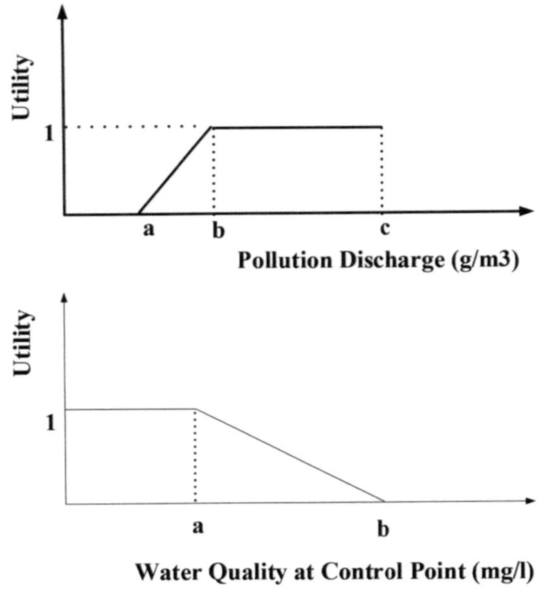

Figure 1.
Utility function of allocated pollution discharge and water quality at control points.

8. Application of bankruptcy method in waste load allocation and water quality management

The equitable and efficient allocation of water among different users is a contentious and complex issue. One analytical approach for resolving resource allocation disputes is bankruptcy theory, traditionally used to distribute assets among creditors when funds are insufficient to satisfy all claims. Over time, various bankruptcy rules have been developed, including Proportional (P), Constrained Equal Loss/Award (CEL/CEA), and Talmud (TAL), which aim to distribute assets fairly while simplifying calculations. These methods differ primarily in their definitions of fairness.

This section adapts the application of four bankruptcy rules in water quality management and waste load allocation. In this application, the "asset" represents the TMDL or river's self-purification capacity, while "claims" correspond to pollution loads introduced by stakeholders. These claims must be managed to preserve the river's self-purification ability to meet TMDL. The purpose is to share reduced pollution loads with point and non-point pollution sources as long as the water quality is within the standard at the water quality monitoring/water withdrawal points [26, 27].

8.1 CEA method

The CEA method allocates pollution share or resources as follows: Initially, each beneficiary receives an equal share, up to the lowest claim, ensuring the total allocation does not exceed the total waste load determined in the TMDL process. Fully satisfied creditors are excluded. The process is repeated with remaining pollution sources until all claims are addressed or the maximum waste load determined at the TMDL process is allocated. Mathematically, the CEA rule is expressed as:

$$WLA_i = Minimum\left(WL_i, \frac{TMDL - MOS}{n}\right) \tag{12}$$

Where $\frac{TMDL - MOS}{n}$ is the waste load allocation to pollution source i based on equal share, WL_i is the maximum desirable pollution discharge (claim) of polluter i, $TMDL - MOS$ is considered as a total asset, and n is the number of pollution sources.

Applying this method to water quality, each polluter's allowable discharge increases gradually until reaching the maximum permissible concentration, resulting in a water quality standard at the water withdrawal/control points. This kind of solution is similar to socialist games. This kind of game is based on the principle of eco-socialism [28].

8.2 CEL method

The CEL method allocates resources as follows: Initially, each beneficiary receives their full claim, which equals the desirable dischargeable waste load for each pollution source. Creditors receiving zero allocation are excluded. Remaining resources are distributed among remaining creditors. Mathematically, the CEL rule is expressed as:

$$WLA_i = Maximum(WL_i - d, 0) \tag{13}$$

$$\sum_{i=1}^{n} WLA_i = TMDL - MOS \tag{14}$$

In the context of water quality, the CEL rule equally reduces pollutant concentrations from each source to achieve the desired pollution concentration, achieving water quality standards at intake/control points.

8.3 Proportional (P)

The P rule allocates resources as follows: A pollution reduction percentage is assigned to each beneficiary, ensuring the total waste load allocation remains within the asset limit, which equals to TMDL-MOS. In water quality management, this percentage is adjusted to achieve the desired water quality standard at the water withdrawal/monitoring point. Mathematically, the P rule is expressed as:

$$P = \left(\frac{TMDL - MOS}{\sum_{i=1}^{n} WL_i}\right) \tag{15}$$

$$WLA_i = p * WL_i \tag{16}$$

This rule proportionally reduces pollutant concentrations from all pollution sources to maintain the required standard for water quality at withdrawal/control sites.

8.4 Talmud (TAL)

The TAL rule combines CEA and CEL: Initially, half of each claim is allocated. If the total claims exceed half the asset, CEA is applied to the halved claims. Otherwise, CEL is applied. The allocation process continues based on the selected method until all claims are addressed. Mathematically, the TAL rule is expressed as:

$$TC = \sum_{i=1}^{n} WL_i \tag{17}$$

$$TAWL = TMDL - MOS \tag{18}$$

$$WLA_i = \begin{cases} CEA(TAWL, 0.5 * TC) & \text{if } TAWL < 0.5TC \\ 0.5 * TC + CEL(TAWL - 0.5 * TC, 0.5 * TC) & \text{if } TAWL > 0.5TC \end{cases} \tag{19}$$

Where TC is total pollution waste load discharge claimed by each polluter and TAWL is maximum dischargeable load based on TMDL study. The TAL rule behaves as either CEA or CEL considering the relation between half of the total claims and the asset. Selecting a suitable bankruptcy method for water resource allocation is challenging due to differing stakeholder preferences. Evaluating the sustainability and acceptability of each method requires careful analysis [29].

9. Application of prisoner's dilemma game to simulate agent response in execution TMDL policies in a river basin water quality management

The Prisoner's dilemma is a traditional and classical game theory model that, faced with a dilemma, two players decide to cooperate or defect. In the original formulation, two suspects are arrested for a crime, but there is insufficient evidence for a conviction. Separated and unable to communicate, each prisoner must decide whether to confess or remain silent. Mutual silence results in minimal punishment for both. If one confesses and the other remains quiet, the confessor is released, and the silent individual receives the harshest penalty. If both confess, they each receive a moderate sentence. The dilemma lies in the uncertainty about the other prisoner's choice, leading to a dominant strategy of betrayal, even though mutual cooperation would yield the best outcome for both [30].

Water resource conflicts can exhibit similar characteristics to the Prisoner's dilemma. Multiple stakeholders face decisions about pollution levels, where individual incentives may conflict with collective benefits. If all parties reduce pollution, water quality improves for everyone. However, if one party pollutes while others cooperate, the polluting party gains an advantage. This situation can lead to a race to the bottom with declining water quality [31, 32].

To overcome the Prisoner's dilemma in water management, strategies such as clear communication, binding contracts, and trust-building are essential. Additionally, introducing regulatory measures, such as monitoring and penalties for non-compliance, can deter defection and encourage cooperation. By modifying the payoff structure through stricter regulations and enforcement, it is possible to transform from a non-cooperative game (dilemma of Prisoners) into a cooperative game where collaboration becomes the dominant strategy. However, implementing such measures requires strong regulatory authority and the ability to enforce penalties effectively [21].

10. Application of chicken game to simulate upstream pollution load and downstream water withdrawal quality, especially in transboundary river basins

The chicken game is one of ancient and classical game theory models involving two individuals driving toward each other on a narrow bridge. The first to swerve (coward out) loses, while the other wins. Neither player desires this outcome, but a head-on collision is even worse. Both players face a dilemma: prioritize pride and risk a collision, or yield and suffer the humiliation of being labeled a "chicken." A tie occurs if both swerve, resulting in no clear winner but preserving pride [33].

Payoffs in the chicken game are determined by the outcome's desirability. The level of payoff significantly influences the preference for the outcome. Unlike the Prisoner's dilemma, where mutual defection leads to a not the best possible result but better than the worst, both prisoners suffer the worst outcome in the chicken game if they both pursue a "free ride."

The chicken game is less common in water quality management literature compared to the Prisoner's Dilemma or coordination games. However, the Afghanistan-Iran-Turkmenistan conflict over the Harrirud River exhibits chicken game characteristics. Upstream dam construction by Afghanistan (Salma dam) reduced water availability for downstream countries, Iran and Turkmenistan. The need for maintenance on shared infrastructure, like the Doosti dam, created a similar dilemma. Both countries desired to minimize costs while maximizing benefits, leading to a standoff. Similar to the original chicken game, the worst outcome was inaction, resulting in severe consequences for all parties [34, 35].

An example of a non-cooperative game in the water quality and quantity management game is the water conflict between Afghanistan-Iran-Turkmenistan on the Harrirud River, which originates from Afghanistan and forms the border between Iran and Turkmenistan and is discharged to Doosti dam. Doosti dam supplies agricultural demand of Iran and Turkmenistan and also supplies drinking water for Mashhad City, which is the second biggest city in Iran. Upstream of Harrirud, Afghanistan constructed the Salma dam and developed agricultural lands downstream. During these years, inflow water quality and quantity to Doosti dam decrease. The conflict between neighboring countries persists, often intensified by Afghanistan's droughts and political turmoil. Throughout this period, the conflict resembled a chicken game. In this way, the Iranian government tries to supply water from other sources and reduce dependency on the domestic water supply of Mashhad City from Doosti dam [36, 37].

To encourage cooperation in chicken game scenarios, increasing the costs of defection is crucial. By imposing penalties for non-cooperation, a governing body can incentivize collaboration. This transforms the game from a chicken game to a cooperative game where mutual benefit is the dominant strategy [38].

11. Application of stag hunt (assurance) game to simulate discharged pollution load to shared sea/lake like Persian Gulf or Caspian Sea water quality

The stag hunt game is a model that two hunters can choose between cooperating to hunt a stag, yielding a higher payoff for both, and individually hunting a rabbit, resulting in a lower payoff. Successful stag hunting requires mutual cooperation, while hare hunting is independent. The worst outcome for an individual is to cooperate while the other defects [39].

The stag hunt shares similarities with the Prisoner's dilemma as both involve a cooperative outcome that is Pareto optimal but risky. However, unlike the dilemma of Prisoners game, the game of stag hunt does not have a completely dominant strategy and has multiple Nash equilibria. Unlike the chicken game, where players pursue opposite strategies, the stag hunt involves players mirroring each other's actions [40].

While the stag hunt might not appear as a dilemma, game theory recognizes the potential for non-cooperation due to a lack of trust, leading to a suboptimal outcome.

Water resource management often resembles a stag hunt. Countries sharing a water body face a choice between cooperating to reduce pollution for mutual environmental benefits and acting independently, prioritizing short-term gains. Similar to the original game, mutual cooperation yields the best outcome, but the risk of one party defecting can deter collaboration [41].

Trust plays a pivotal role in the stag hunt. Without trust, players may opt for the risk-free but less optimal strategy of individual action. Unlike the Prisoner's dilemma, where free-riding is tempting, the stag hunt encourages cooperation when signs of cooperation are observed. However, the absence of repeated interactions can hinder trust-building [42].

To facilitate cooperation in stag hunt-like situations, clear communication, negotiation, and potentially external enforcement mechanisms are essential. Regional agreements and organizations, such as the Regional Organization for the Protection of the Marine Environment (ROPME), can provide frameworks for cooperation and address shared challenges.

12. Case studies and examples

In this section, an example of scenario analysis based on game theory concept and a case study are presented.

12.1 Non-cooperative decision-making example

Assume a game for simulation of the water polluters (industrial, agricultural, and urban waste water discharge) and Department of Environment (DoE) decisions for pollution reduction and improvement of wastewater treatment efficiency. Assume that the pollution load of each polluter varies 20% more or less than that of the mean historical pollution data. The treatment efficiency increase is 55%, which leads to a reduction of pollution load discharged (quality and quantity) to the aquifer.

The two players (DoE and polluters) state are presented in **Table 1**. The DoE can help and invest for wastewater treatment efficiency improvement or do not pay any loan. When DoE pays a loan, the treated wastewater can be discharged to an aquifer in order to reduce groundwater table drawdown.

States	Department of Environment	Polluters	State
(Pay, Do not Accept)	The DoE will pay all the costs of wastewater treatment plants and recharge the aquifer.	Polluters invest them self and improve treatment efficiency but sell or reuse treated wastewater.	The groundwater quality will improve but due to recharge reduction groundwater table drawdown.
(Pay, Accept)	The DoE will pay all the costs of wastewater treatment plants.	Polluters will improve treatment efficiency by governmental loan.	Indicating cooperative relation between DoE and polluters and the aquifer recharged with treated wastewater, there is less pollution and no drawdown.
(Do not Pay, Do not Accept)	The DoE do not pay any loan and pollution discharge will continue.	Polluters do not improve treatment efficiency or improve the efficiency and sell the treated water.	The current condition will continue or the groundwater quality will improve but due to recharge reduction groundwater table drawdown.
(Do not Pay, Accept)	The DoE do not pay any loan and pollution discharge will continue.	Polluters will improve treatment efficiency and discharge the treated wastewater to the aquifer.	This state will not accrue in reality. In few cases environmental awareness will help to have this solution.

Table 1.
Descriptions of department of environment and water polluter's strategies in the wastewater treatment construction game.

Polluters have two alternatives: either to accept the investment by DoE (Accept) or invest by themselves and sell the treated wastewater or reuse it (Do not Accept).

Features and definitions of each player's decision state are presented in **Table 1**. **Figure 2** shows a strategic matrix of the efficiency improvement game between the DoE and polluters. In this table, (a, b) shows the player payoffs; a: indicates the DoE payoff, and b: indicates polluters' payoff. When the polluters invest, the groundwater pollution may reduce, but due to discharge reduction, the groundwater table may drawdown.

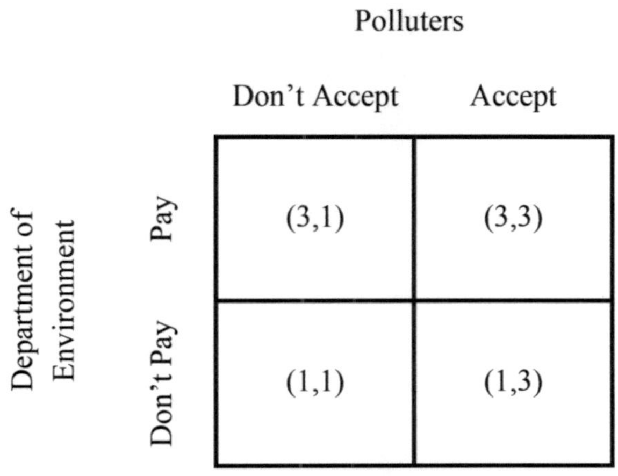

Figure 2.
Matrix form of efficiency improvement game.

Table 2 shows different states, including Nash, GMR, SMR, and SEQ stables between two players. The state (Do not Pay, Accept) represents the scenario in which the DoE does not take any action, but polluters reduce their pollution discharge. Since polluters have no desire to reduce pollution without any DoE support, this scenario will not come to reality.

On the other hand, the state (Pay, Accept) is a stable solution in all stability definitions except for Nash. This state can be achieved by the cooperative between government and farmers. This stated is not a stable state from the Nash definition because of the low foresight of the Nash definition.

Table 2 shows stables for polluters and DoE in the efficiency improvement of wastewater treatment game.

12.2 Case study: Application of bankruptcy method in river water quality management

In this study, a novel approach to managing water quality is introduced to address conflicts among river system beneficiaries through the application of bankruptcy rules.

These rules have been adapted to align with water quality parameters. Essentially, the river water quality index is treated as the "asset," while the pollutants discharged by various sources are considered the "claims" of the stakeholders. The key parties involved in this issue are the Environmental Protection Agency (EPA) and pollution sources, including both point and non-point sources. The research incorporates four widely-used bankruptcy rules, specifically tailored for water resource allocation, to distribute pollution responsibilities fairly among polluters, aiming to enhance overall river water quality. To achieve this objective, a simulation-optimization model is employed, integrating QUAL2Kw with Particle Swarm Optimization (PSO). Here, PSO is utilized to optimize waste load allocation (WLA) using bankruptcy methods, with a focus on minimizing dissolved oxygen (DO) violations downstream from the regulatory standard (**Figure 3**).

The simulation-optimization model was applied based on the lowest October river inflow (as indicated in **Table 3**). The CEA method allocated the minimum

Stability definition	Nash		GMR (general meta-rationality)		SMR (symmetric meta-rationality)		SEQ (sequential stability)	
	Polluters	DoE	Polluters	DoE	Polluters	DoE	Polluters	DoE
(Pay, Do not Accept)		✓		✓		✓		✓
(Pay, Accept)			✓	✓	✓	✓	✓	✓
(Do not Pay, Do not Accept)	✓		✓		✓		✓	
(Do not Pay, Accept)	✓	✓	✓	✓	✓	✓	✓	✓

Table 2.
Different types of stables for DoE and polluters in the wastewater treatment construction game.

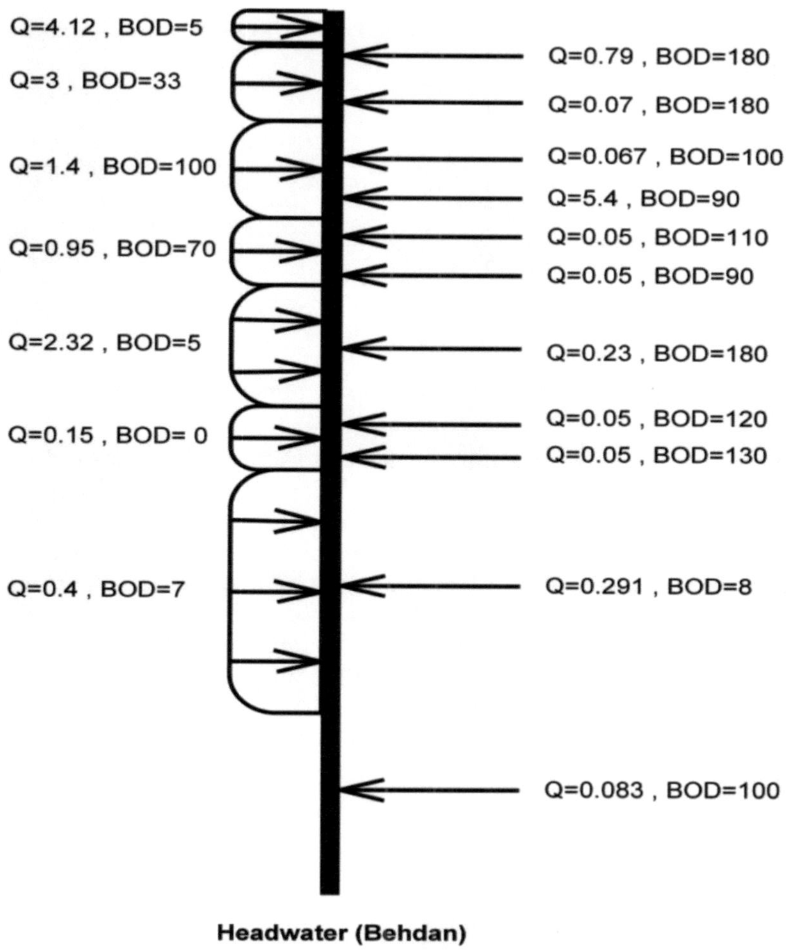

Figure 3.
Scheme of point and non-point source pollutions discharged to Zarjub River, Gillan Province, Iran [26].

Source type	Number	DO	Initial BOD5	CEA	CEL	P	TAL
Point source	1	8.0	100	4.97	0.87	5	1.97
	2	7.32	8	4.97	0	0.4	1.97
	3	3.0	130	4.97	30.87	6.5	1.97
	4	3.0	120	4.97	20.87	6	1.97
	5	3.0	180	4.97	80.87	9	1.97
	6	5.0	90	4.97	0	4.5	1.97
	7	0.1	110	4.97	10.87	5.5	1.97
	8	0.1	90	4.97	0	4.5	1.97
	9	0.1	100	4.97	0.87	5	1.97
	10	0.1	180	4.97	80.87	9	1.97
	11	0.1	180	4.97	80.87	9	1.97

Source type	Number	DO	Initial BOD5	CEA	CEL	P	TAL
Non-point source	1	9	7	4.97	0	0.35	1.97
	2	8	0	0	0	0	0
	3	3	5	4.97	0	0.25	1.97
	4	0	70	4.97	0	3.5	1.97
	5	0	100	4.97	0.87	5	1.97
	6	0	33	4.97	0	1.65	1.97
	7	0.5	5	4.97	0	0.25	1.97

Table 3.
Discharged BOD5 (mg/l) from pollutants based on bankruptcy in a deterministic condition [25].

concentration, ranging from the initial value up to 4.97 mg/l, to the dischargers. Since all initial contamination levels were below 4.97 mg/l, dischargers were only allowed to release concentrations of 4.97 mg/l to ensure the DO level at the checkpoint was maintained. Clearly, this approach primarily benefits those with the lowest pollution levels. The CEL method reduced concentrations by 99.13 mg/l from the sources but had no impact on those whose concentrations were already below this threshold. This rule thus favors the major contributors to contamination. The P method set a fixed allocation rate of 0.05% for all contamination sources, whether point or non-point. Lastly, the TAL method, similar to the CEA, adjusted to the river conditions by limiting discharge to either zero or half of the initial concentration.

13. Conclusion

Game theory offers a valuable framework for understanding and addressing complex water resource conflicts involving multiple stakeholders and criteria. This approach can effectively capture the intricate interplay of engineering, socioeconomic, and political factors without requiring detailed quantitative data or relying solely on traditional financial, economic, or physical metrics. Game theory can define the feasibility of optimal solutions and illuminate the behavior of stakeholders in certain situations. While the actions of decision-makers may appear irrational from a systems engineering perspective, game theory reveals how rational self-interest can lead to suboptimal outcomes for the overall system.

Through simplified two-player water resource games, this study demonstrated how game theory can expose suboptimal solutions and the influence of self-interest on decision-making. While these examples are basic, they provide foundational insights for analyzing more complex real-world scenarios. Non-cooperative game theory is particularly useful for modeling conflicts with limited quantitative information, relying instead on ordinal data. This flexibility offers a significant advantage over classical system analysis and optimization methods.

While many water disputes have been analyzed as Prisoner's dilemmas, the analysis of different game structures (chicken and stag hunt) reveals the diversity of water resource challenges. The potential for third-party intervention to modify game structures and encourage cooperation was also explored. Understanding how game dynamics evolve over time is crucial for accurately predicting stakeholder behavior.

The example of a deteriorating water quality situation due to short-term self-optimization highlights the importance of considering long-term consequences.

To promote environmentally friendly practices, several strategies are proposed. These include strengthening government regulation, implementing reward and punishment systems for regulators, providing incentives for enterprises, enhancing interagency cooperation, and fostering public awareness. By addressing information asymmetries, improving regulatory efficiency, and encouraging public participation, a more robust environmental protection framework can be established.

Acknowledgements

The author acknowledges the use of ChatGPT 4.0 and Google Gemini for language polishing of the manuscript.

Author details

Ali Moridi
Civil, Water and Environmental Engineering Faculty, Shahid Beheshti University, Tehran, Iran

*Address all correspondence to: a_moridi@sbu.ac.ir

References

[1] Karamouz M, Szidarovszky F, Zahraie B. Water Resources Systems Analysis. NewYork: CRC Press; 2003

[2] Zare Farjoudi S, Moridi A, Sarang A. Multi-objective waste load allocation in river system under inflow uncertainty. International journal of Environmental Science and Technology. 2021;**18**:1549-1560

[3] Kerachian R, Karamouz M. Optimal reservoir operation considering the water quality issues: A stochastic conflict resolution approach. Water Resources Research. 2006;**42**(12):12401-12417

[4] Lahti A. Biodiversity-related negotiations in city planning: A role-playing game. Master's thesis. Finland University of Helsinki; 2022

[5] Jiang K, You D, Li Z, Shi S. A differential game approach to dynamic optimal control strategies for watershed pollution across regional boundaries under eco-compensation criterion. Ecological Indicators. 2019;**105**:229-241

[6] Liu B, Huang JJ, McBean E, Li Y. Risk assessment of hybrid rain harvesting system and other small drinking water supply systems by game theory and fuzzy logic modeling. Science of the Total Environment. 2020;**708**:134436

[7] Aghasian K, Moridi A, Mirbagheri A, Abbaspour M. Selective withdrawal optimization in a multipurpose water use reservoir. International journal of Environmental Science and Technology. 2019;**16**:5559-5568

[8] Kerachian R, Karamouz M. A stochastic conflict resolution model for water quality management in reservoir–river systems. Advances in Water Resources. 2007;**30**(4):866-882

[9] Yahyaee AR, Moridi A, Sarang A. A new optimized model to control eutrophication in multi-purpose reservoirs. International journal of Environmental Science and Technology. 2021;**18**:3357-3370

[10] Karamouz M, Moridi A, Fayazi HM. Dealing with Conflict over Water Quality and Quantity Allocation: A Case Study. Vol. 15. no. 1. Tehran, Iran: Scientia Iranica; 2008. pp. 34-49

[11] Loucks DP, Van Beek E. Water Resource Systems Planning and Management: An Introduction to Methods, Models, and Applications. Switzerland: Springer; 2017

[12] Shi G-M, Wang J-N, Zhang B, Zhang Z, Zhang Y-L. Pollution control costs of a transboundary river basin: Empirical tests of the fairness and stability of cost allocation mechanisms using game theory. Journal of Environmental Management. 2016;**177**:145-152

[13] Karamouz M, Moridi A. Risk based conflict resolution model for river water quality management. In: World Environmental and Water Resource Congress 2006: Examining the Confluence of Environmental and Water Concerns. Omaha, Nebraska, United States; 21 May 2006. pp. 1-8

[14] Karamouz M, Akhbari M, Moridi A. Resolving disputes over reservoir-river operation. Journal of Irrigation and Drainage Engineering. 2011;**137**(5):327-339

[15] Tayia A. Transboundary water conflict resolution mechanisms:

Substitutes or complements. Water (Basel). 2019;**11**(7):1337

[16] Mianabadi H, Mostert E, van de Giesen N. Trans-boundary river basin management: Factors influencing the success or failure of international agreements. In: Hipel K, Fang L, Cullmann J, Bristow M, editors. Conflict Resolution in Water Resources and Environmental Management. Cham: Springer; 2015. pp. 133-143

[17] Tu G, Yu C, Yu F. Cooperative strategies in transboundary water pollution control: A differential game approach. Water (Basel). 2024;**16**(22):3239

[18] Yang Z, Zeng Y, Cai Y, Tan Q. An integrated game-theory based model for trans-boundary water resources management in North China: A case study in the guanting reservoir basin (grb), Beijing. International Journal of Software Engineering and Knowledge Engineering. 2008;**18**(04):461-483

[19] Yang Y, Liu Y, Dai J, Zeng Y. Cost-sharing mechanism of water pollution control in main and subbasins based on Stackelberg game model. Mathematical Problems in Engineering. 2022;**2022**(1):6559840

[20] Tyagi A, Shortle JS. Modeling Endogenous Change in Water Allocation Mechanisms: A Non-Cooperative Bargaining Approach. In: Proceedings of the 2016 Annual Meeting—Agricultural and Applied Economics Association (AAEA) Conferences, Boston, MA, USA. 31 July–2 August 2016. p. 28

[21] Madani K, Pierce TW, Mirchi A. Serious games on environmental management. Sustainable Cities and Society. 2017;**29**:1-11

[22] Yang Y, Liu Y, Yuan Z, Dai J, Zeng Y, Khan MYA. Analyzing the water pollution control cost-sharing mechanism in the Yellow River and its two tributaries in the context of regional differences. Water (Basel). 2022;**14**(11):1678

[23] Madani K, Dinar A. Non-cooperative institutions for sustainable common pool resource management: Application to groundwater. Ecological Economics. 2012;**74**:34-45

[24] Eyckmans J. Nash implementation of a proportional solution to international pollution control problems. Journal of Environmental Economics and Management. 1997;**33**(3):314-330

[25] Aghasian K, Moridi A, Mirbagheri A, Abbaspour M. A conflict resolution method for waste load reallocation in river systems. International Journal of Environmental Science and Technology. 2019;**16**:79-88

[26] Moridi A. A bankruptcy method for pollution load reallocation in river systems. Journal of Hydroinformatics. 2019;**21**(1):45-55

[27] Farjoudi SZ, Moridi A, Sarang A, Lence BJ. Application of probabilistic bankruptcy method in river water quality management. International Journal of Environmental Science and Technology. 2021;**18**:3043-3060

[28] Rezaei M, Mousavi S-F, Moridi A, Eshaghi M, Karami H. An optimal socialist cooperative game theory model in agricultural sector (case study: Dezful-Andimeshk Plain). International Journal of Nonlinear Analysis and Applications. 2021;**12**(Special Issue):719-731. DOI: 10.22075/ijnaa.2021.23423.2539

[29] Mehrparvar M, Ahmadi A, Safavi HR. Social resolution of conflicts over water resources allocation in a river

basin using cooperative game theory approaches: A case study. International Journal of River Basin Management. 2016;**14**(1):33-45

[30] Furber A, Medema W, Adamowski J. Assessing the benefits of serious games to support sustainable decision-making for transboundary watershed governance. Canadian Water Resources Journal/Revue Canadienne des Ressources Hydriques. 2018;**43**(4):401-415

[31] Madani K. Game theory and water resources. Journal of Hydrology. 2010;**381**(3-4):225-238

[32] Madani K, Hipel KW. Non-cooperative stability definitions for strategic analysis of generic water resources conflicts. Water Resources Management. 2011;**25**:1949-1977

[33] Yang T, Liao H, Du Y. A dynamic game modeling on air pollution mitigation with regional cooperation and noncooperation. Integrated Environmental Assessment and Management. 2023;**19**(6):1555-1569

[34] Madani K, Dinar A. Cooperative institutions for sustainable common pool resource management: Application to groundwater. Water Resources Research. 2012;**48**(9):9553-9568

[35] Jeong H, Ha S, Kim H. Flood protection cost allocation using cooperative game theory for adapting infrastructure to climate change. Journal of Water Resources Planning and Management. 2018;**144**(4):04018010

[36] Nazari Mejdar H, Moridi A, Najjar-Ghabel S. Water quantity–quality assessment in the transboundary river basin under climate change: A case study. Journal of Water and Climate Change. 2023;**14**(12):4747-4762

[37] Moridi A. Dealing with reservoir eutrophication in a trans-boundary river. International Journal of Environmental Science and Technology. 2019;**16**:2951-2960

[38] Tulkens H. Economics, Game Theory and International Environmental Agreements: The Ca'Foscari Lectures. Vol. 7. New Jersey, USA: World Scientific; 2019

[39] Kicsiny R, Varga Z. Differential game model with discretized solution for the use of limited water resources. Journal of Hydrology. 2019;**569**:637-646

[40] Di D, Wu Z, Wang H, Lv C. A double-layer dynamic differential game model for the optimal trading quantity of water and price setting in water rights transactions. Water Resources Management. 2020;**34**:245-262

[41] Hemati H, Abrishamchi A. Water allocation using game theory under climate change impact (case study: Zarinehrood). Journal of Water and Climate Change. 2021;**12**(3):759-771

[42] Avarideh F, Attari J, Moridi A. Modelling equitable and reasonable water sharing in transboundary rivers: The case of Sirwan-Diyala river. Water Resources Management. 2017;**31**:1191-1207

Chapter 6

Development of a Game Theoretical Bankruptcy Methods for Shared Water Resource Allocation

Reza Javidi Sabbaghian and Mohammad Ali Tolouei Virani

Abstract

The study explores the application of bankruptcy theory within the framework of cooperative game theory to address conflicts over shared water resources. It emphasizes the advantage of game theory, particularly bankruptcy methods, over traditional optimization techniques for resolving these complex disputes. The study focusses on the usage of bankruptcy methods in situations where water resources are bankrupt and must be shared among multiple beneficiaries, highlighting its effectiveness in achieving equitable outcomes. The study examines several bankruptcy methods, including proportional (PRO), adjusted proportional (AP), constrained equal awards (CEA), constrained equal losses (CEL), Talmud (TAL), Piniles (P), modified constrained equal losses (MCEL), and groundwater-based (GW) methods. Through case studies in regions such as Asia, Europe, Africa, and North America, the research shows the practical application of these methods in fostering long-term cooperation and reducing water stresses. The study concludes that adopting game theoretical bankruptcy methods can lead to more equitable water allocation strategies among multiple stakeholders.

Keywords: game theory, bankruptcy methods, shared water resources, water allocation, stakeholders

1. Introduction

1.1 Global water distribution

Approximately 71% of the Earth's surface is covered by oceans that hold 97.5% of the planet's water resources. However, only 2.5% of this water is freshwater, with about 68.7% of it locked in ice caps and glaciers. The remaining 30.1% of freshwater, which sustains all terrestrial life, is mostly stored underground as groundwater, with only a small portion found in rivers, lakes, and other surface water sources [1]. As the global population rapidly increases, societies face serious challenges in meeting the rising demands for infrastructure, agriculture, environmental conservation, food production, and water supply [2]. Water resources are critical for food and energy production, social and economic sustainability, ecosystem integrity, biodiversity, and essential ecosystem services [3].

1.2 Water security

Billions of people worldwide, especially in rural areas and least developed countries (LDCs), still live without sufficient high-quality water, which is essential for human survival [4, 5]. Currently, over 0.61 billion people (out of 7.78 billion across 186 countries) face critical water insecurity. According to the United Nations (UN) assessments, water security for each country has been evaluated using a score ranging from 0 to 100, where 0 indicates critically insecure, and 100 indicates moderately secure, as shown in **Figure 1** [5].

In recent years, the security of our limited freshwater resources has been increasingly threatened by the competing demands of growing populations and global economies. Accordingly, water security is defined as the ability of a population to ensure sustainable access to adequate quantities of high-quality water for supporting livelihoods, human welfare, and socioeconomic development. This also includes safeguarding against water pollution and water-related disasters, while preserving ecosystems and maintaining a peaceful and sustainable political environment [5].

1.3 Shared water resources

Among the limited freshwater resources, certain dispersed sources are collectively utilized by humans and are known as shared water resources. These include rivers and lakes as surface water resources, and aquifers as groundwater resources, which cross or are shared by multiple political, economic, or social boundaries [6]. Shared water resources account for 60% of the world's freshwater flows. A total of 153 countries have territory within at least one of the 286 shared river and lake basins, as well as within 592 shared aquifer systems [4]. However, these critical resources face growing challenges due to population increase, unsustainable development practices, and

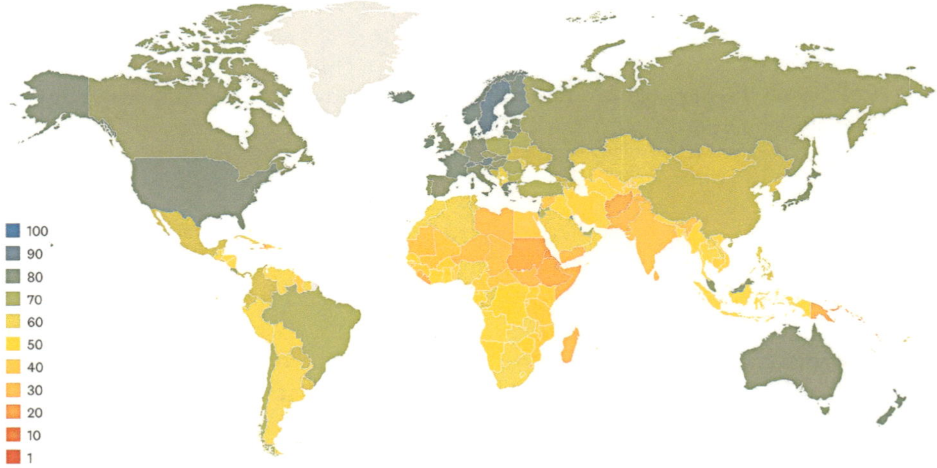

Figure 1.
Distribution of national water security scores ranging from 0 (critically insecure) to 100 (moderately secure) worldwide [5].

climate change. As a result, 80% of the global population is at serious risk of water insecurity [7]. Therefore, cooperation among stakeholders on shared water resources is necessary for effective management, as it can bring various advantages to participating stakeholders the ability for bringing various advantages to participating stakeholders [8].

1.4 Allocation strategies

It is important to adopt effective approaches to reduce conflicts and ensure the long-term sustainable development of water resources [3, 9]. Sustainable development aims to balance economic growth, environmental concerns, and social welfare needs, meeting present and future demands without compromising the resources available for future generations [10]. To assist stakeholders and policymakers in the equitable allocation of water resources, extensive research has been conducted. In this regard, the related research has focused on two main approaches: optimization techniques and the application of game theory [10, 11].

1.4.1 Optimization techniques

Optimization techniques can generally be described as the process of maximizing or minimizing objective function(s) within a defined decision space and subject to a set of constraints [12]. In cases such as the allocation of shared water resources, where multiple stakeholders with competing interests try to maximize their own demands, considering various strategies in mathematical modeling can lead to increased synchronization errors. This can result in outcomes that are not only unrealistic but also impractical for the stakeholders involved [13].

1.4.2 Game theory

Unlike optimization techniques, which often require simplifying complex interactions into the form of objective function(s), game theory models these interactions in their full complexity, accounting for each stakeholder's strategies and payoffs. Consequently, researchers have increasingly turned to game theory as a complementary approach to address the limitations of optimization techniques in water resource management [14, 15]. Game theory, as a study of strategic interactions among rational decision-makers, offers a robust framework for analyzing and predicting the behavior of multiple stakeholders with conflicting interests [13]. In water resource management, game theory addresses issues such as quantitative surface and groundwater management, water quality management, water allocation, and the sharing of water among beneficiaries [16]. By applying game theory, we can enhance resource allocation efficiency and sustainability, and foster stable long-term cooperation among stakeholders [13]. The practical application of game theory in economics began with Neumann and Morgenstern in 1944, who summarized its mathematical foundations and economic significance [17]. Overall, game theory offers essential insights and methodologies for tackling intricate water resource issues, enhancing both efficiency and collaborative efforts.

Nash revolutionized game theory by introducing the Nash equilibrium for n-player games [10]. Before Nash's groundbreaking work, game theory primarily focused on zero-sum games, where one player's gain was exactly balanced by another's loss [18]. The Nash equilibrium, which applies to nonzero-sum games, expanded the field by allowing the analysis of a wider range of strategic scenarios. This advancement made game theory applicable to real-world situations where mutual gains are possible. Nash's pioneering research laid the groundwork for applying game theory across various fields, including water resources management [13, 19]. In this domain, game theory helps optimize the allocation of water among competing stakeholders, resolve conflicts, and ensure sustainable usage. Since then, the application of game theory in water resources planning and management has evolved, incorporating diverse concepts and methodologies to address challenges related to resource sharing and conflict resolution.

Game theory is divided into two main categories: cooperative and noncooperative [16]. Noncooperative game theory examines strategies where players act independently, often with competing interests, and make decisions without collaboration [13]. In contrast, cooperative game theory explores situations where groups or coalitions of players work together to achieve objectives and share the benefits of their cooperation. The main focus in cooperative games is the allocation of collective gains among the coalition members. Research into the application of game theory in water resources management has shown a preference for cooperative game theory among researchers [13]. This preference arises from the nature of water resource issues that typically involve multiple stakeholders with interconnected interests, necessitating joint decision-making.

1.5 Bankruptcy concept

In shared water resource management, a primary challenge arises when the available water supply is insufficient to meet the diverse demands of all beneficiaries within a basin [20]. This situation closely resembles bankruptcy problems in economics, where the assets of a bankrupt entity are inadequate to cover all outstanding claims. In both contexts, decision-makers must navigate competing interests and limited resources to develop fair and justifiable solutions. Applying bankruptcy theory, as a subcategory of game theory, can enhance the robustness and fairness of water allocation policies, contributing to the long-term sustainability of water resources.

2. Methodology

In this section, we explore the application of bankruptcy theory within the framework of game theory to address conflicts in the allocation of shared water resources. We specifically outline various bankruptcy methods and explain how their principles can be applied to interactions among stakeholders, considering their respective claims and the available resources. These methods allocate resources by evaluating each stakeholder's claims relative to the total available resources.

2.1 Bankruptcy theory

In water resource allocation under bankruptcy conditions, the objective is to share a limited set of resources among multiple stakeholders in an equitable and efficient manner. Let E represents the total available resource, and c_i denotes the claim of stakeholder i. If $\sum_{i=1}^{N} c_i > E$, it indicates bankruptcy conditions, meaning the total claims exceed the available resource. Under these conditions, the limited resource should be shared among $N \geq 2$ stakeholders, who may have varying relative weights. Each stakeholder has claims represented by $c = (c_1, \dots, c_n)$, where each claim c_i should be nonnegative $(c_i \geq 0)$. Additionally, they have contributions represented by $a = (a_1, \dots, a_n)$, where each a_i should also be nonnegative $(a_i \geq 0)$. The objective of a bankruptcy method is to determine the appropriate allocation for each stakeholder, denoted by $x_i = \psi_i(E, c, a) \geq 0$. The allocation should meet the following criteria [20]:

1. Total resource constraint: The sum of the allocations should not exceed the total available resource, and the entire available resource must be fully allocated. Mathematically, this is expressed as $(\sum_{i=1}^{n} x_i = \sum_{i=1}^{n} a_i = E)$ for all $i \in N$.

2. Nonnegativity and claim limits: Each stakeholder should receive a nonnegative quota that does not exceed their claim $(0 \leq x_i \leq c_i)$.

3. Equity for equal claims: Stakeholders with identical claims should receive the same allocation if $(c_i = c_j, then\ x_i = x_j)$. Additionally, the allocations should be positively correlated with the claims, meaning that if $(c_i \leq c_j, then\ x_i \leq x_j)$. This ensures that the allocation reflects the relative size of each stakeholder's claim while maintaining fairness and proportionality across different claims.

By adhering to these criteria, the allocation process ensures that the total quotas precisely match the available resources. To achieve an equitable and efficient allocation of limited shared resources under bankruptcy conditions, various bankruptcy methods can be employed. To bridge the gap between theoretical concepts and practical application, we present a flowchart in **Figure 2** that illustrates the relationship between bankruptcy methods and game theory approaches in the context of shared water resource allocation conflicts.

As shown in **Figure 2**, bankruptcy problems are a specialized area within cooperative game theory (CGT) focuses on the equitable distribution of assets among multiple stakeholders [20, 22]. This approach aims to balance the interests of all stakeholders involved, ensuring a fair outcome in bankruptcy scenarios.

2.2 Bankruptcy methods

In **Figure 2**, various bankruptcy methods for allocating shared water resources are categorized into three groups based on their underlying principles. The first category, asset-based methods, includes approaches that allocate resources only based on the total amount of resources available to be shared among stakeholders. The second category, deficit-based methods, comprises approaches that determine allocations primarily based on the level of resource deficiency, prioritizing the relative deficit among stakeholders. The third category, asset-deficit-based methods, integrates both the total amount of resources and the deficiency levels into the allocation process.

This classification framework enhances researchers' ability to evaluate and compare different allocation methods for shared water resources. By aligning methods with their foundational principles, this categorization facilitates a more coherent understanding and assessment of their efficiency. In the subsequent sections, we will introduce and review the various methods illustrated in **Figure 2**.

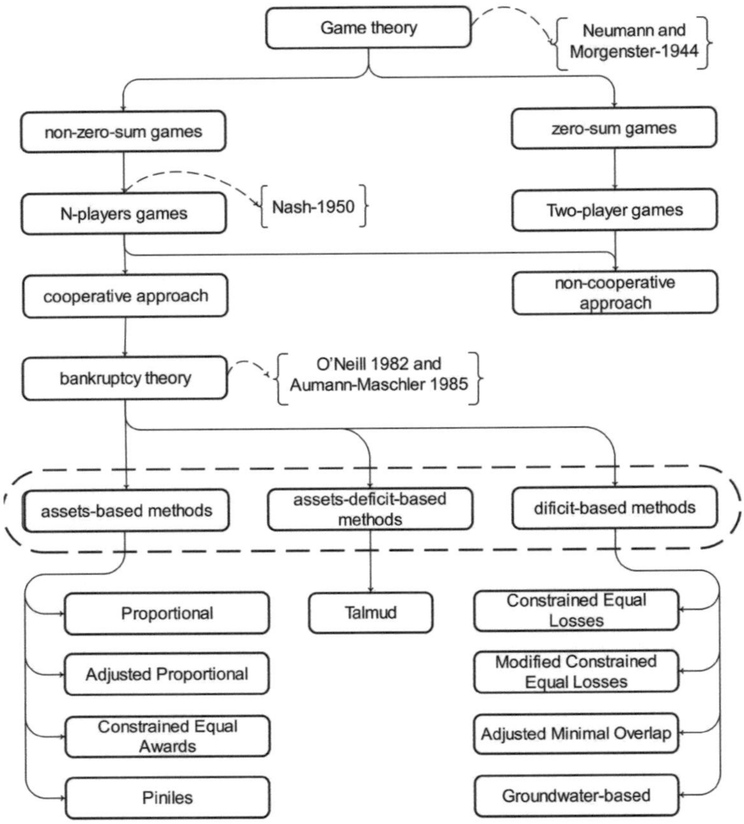

Figure 2.
Relation between the game theory approaches and the bankruptcy methods in water resource management [21].

2.2.1 Proportional (P)

This method establishes a fair approach to distributing water resources by assigning shares that correspond to each stakeholder's claims. It specifies a proportion (λ) that ensures each stakeholder's allocation aligns with their individual claim [23].

$$x_i^{pro} = \left[\rho_{pro}(c_i)\right] \tag{1}$$

$$\rho_{pro} = \left[\left(\frac{E}{\sum_{i=1}^{n} c_i}\right)\right] \tag{2}$$

By this method in Eq. (1), x_i^{pro} represents the quota of resources for stakeholder i, and ρ_{pro} is the proportional factor. In Eq. (2), c_i denotes the claim of stakeholder i, and E is the total amount of resources. In this allocation method, the amount of resources assigned to each stakeholder is determined by applying a proportional factor to their individual claim.

2.2.2 Adjusted proportional (AP)

This allocation method follows a proportional approach. Initially, it determines the quota for each stakeholder by calculating the total claims of all stakeholders, excluding stakeholder i, and subtracting this sum from the total available resources. If resources remain after this deduction, stakeholder i receives a quota equal to the remaining amount. Next, the claim of stakeholder i is adjusted to be the lesser of the remaining resources or the difference between the initial claim and the quota received. After this adjustment, the remaining resources are distributed among the stakeholders according to the principles of proportionality, using the modified claim values. However, if the total claims of all stakeholders, excluding stakeholder i, meet or exceed the total available resources, leaving no surplus, stakeholder i does not receive any initial quota. In this case, the entire resource is allocated based on the proportional principles applied to the initial claims [23].

$$m_i = Max\left[0, E - \sum_{j \neq i} c_j\right] \tag{3}$$

$$c_i' = Min\left[(c_i - m_i), \left(E - \sum\nolimits_{i=1}^{N} m_i\right)\right] \tag{4}$$

In Eq. (3), m_i represents the initial quota for stakeholder i, and E represents the total amount of available resources. In Eq. (4), c_i' denotes the revised claim of stakeholder i after the initial quota is considered, and c_i is the initial claim of stakeholder i.

2.2.3 Constrained equal award (CEA)

The CEA method sets a uniform cap on the maximum amount of resources any single stakeholder can receive. This cap ensures that no stakeholder is allocated more than a specified maximum amount, while also considering their individual claims. The approach guarantees that the sum of all allocated resources equals the total amount available. This makes the CEA method particularly favorable for stakeholders with lower claims compared to others [23].

$$\lambda_{CEA} = \left(\frac{E}{N}\right) \tag{5}$$

$$E = \sum\nolimits_{i=1}^{N} Min[\lambda_{CEA}, c_i] \tag{6}$$

$$x_i^{CEA} = Min[\lambda_{CEA}, c_i] \tag{7}$$

Based on Eq. (7), two primary parameters are crucial: the cap (λ_{CEA}), determined by Eq. (5), where N is the number of stakeholders, and the total amount of resources (E). The condition of Eq. (6) ensures that all resources are allocated without surplus or shortfall, maintaining a balanced distribution while adhering to the cap on individual allocations.

2.2.4 Constrained equal losses (CEL)

The CEL method is a resource allocation approach designed to equitably distribute reductions among stakeholders. As shown in Eq. (7), this method incorporates the inverse principles of the CEA method. The CEL method is particularly advantageous for stakeholders with higher claims, ensuring that their needs are more effectively addressed in the allocation process.

The core idea behind the CEL method is to minimize the maximum loss experienced by any stakeholder relative to their claim. It operates by calculating each stakeholder's quota based on their claim and the total resources available, while imposing a constraint on the maximum allowable loss [23].

$$\lambda_{CEL} = \left(\frac{C - E}{N}\right); 0 \leq E \leq C \tag{8}$$

$$E = \sum\nolimits_{i=1}^{N} Max[(c_i - \lambda_{CEL}), 0] \tag{9}$$

$$x_i^{CEL} = Max[(c_i - \lambda_{CEL}), 0] \tag{10}$$

By ensuring that losses are constrained and balanced across all stakeholders, this method addresses the challenge of resource scarcity. In Eq. (10), x_i^{CEL} represents the quota of resources for stakeholder i, and c_i is the claim of stakeholder i. In Eq. (8), N is the number of stakeholders, λ_{CEL} is a threshold value calculated to ensure that the losses are distributed evenly, C denotes the total claim of stakeholders, and E is the total amount of resources. Constraint on E ensures that the total resources are non-negative and do not exceed the total claim of stakeholders. Eq. (9) ensures that the sum of the assigned quotas equals to the expected total resources.

2.2.5 Modified constrained equal losses (MCEL)

This method refines the principle of CEL method by focusing on the allocation of resources among stakeholders based on their contributions to resource creation. Unlike the CEL method, which distributes total deficit equally, this modified method adjusts the overall deficit by considering each stakeholder's quota in resource creation, as shown in Eqs. (11) and (12) [23]. Stakeholders with a greater role in creating the resource will incur a proportionally smaller share of the total deficit, whereas those with a smaller contribution will bear a larger share of the total deficit.

$$D = C - E \tag{11}$$

$$d_i = \psi_i(\mathrm{N}, \mathrm{E}, c_i, a_i) = \left[\left(\frac{1 - \left(\frac{a_i}{\sum a_i}\right)}{n - 1}\right) * D\right] \tag{12}$$

$$x_i^{MCEL} = c_i - d_i; 0 \leq x_i \leq c_i; \sum_{i \in N} x_i^{MCEL} = E \tag{13}$$

In Eq. (11), the total deficit (D) is determined by subtracting the total amount of resources (E) from the total claim of the stakeholders (C). In Eq. (12), the deficit factor for each stakeholder is based on the inverse effect of that stakeholder's quota in creating the resource. Here, a_i represents the amount of contribution made by the stakeholder to create the resource, and N denotes the total number of stakeholders involved. In Eq. (13), x_i^{MCEL} represents the quota of resources for stakeholder i, and each allocation must be positive and cannot exceed the individual claim of the stakeholder it is assigned to. Furthermore, the total sum of all allocations across stakeholders must precisely match the overall amount of resources available for allocation.

2.2.6 Talmud (TAL)

The Talmud method for distributing resources among stakeholders, developed by Aumann and Maschler in 1985, is based on the concept of half-claims [24]. The core idea is to assess whether the total resource (E) to be distributed is greater or less than half of the total claims. If E exceeds half of the total claims, the method ensures that no one loses more than half of their claim by using the CEL method based on Eq. (14). Conversely, if E is less than half of the total claims, the method ensures that no one receives more than half of their claim, following the CEA method based on Eq. (15) [25].

$$If\left[E \geq \frac{1}{2}C\right] Then\ x_i^{TAL} = \frac{1}{2}c_i + CEL\left[\frac{1}{2}c_i, \left[E - \left(\frac{1}{2}C\right)\right]\right] \tag{14}$$

$$If\left[E \leq \frac{1}{2}C\right] Then\ x_i^{TAL} = CEA\left[\frac{1}{2}c_i, E\right] \tag{15}$$

$$\sum\nolimits_{i=1}^{N} x_i^{TAL} = E \tag{16}$$

In Eqs. (14) and (15), C represents the total claim of the stakeholders, x_i^{TAL} denotes the asssigned quota of stakeholder i by this method, and c_i is the claim of stakeholder i. Eq. (16) ensures that the total distributed quotas sum up to the total resource (E).

2.2.7 Piniles (PIN)

The principles of this method are very similar to those of the Talmud method, with a minor adjustment in calculations. Specifically, if the total resources exceed half of the total claims of the stakeholders, the CEA method is employed instead of the CEL method for further calculations.

$$If\left[E > \frac{1}{2}C\right] Then\ x_i^{PIN} = \frac{1}{2}c_i + CEA\left[\frac{1}{2}c_i, \left[E - \left(\frac{1}{2}C\right)\right]\right] \tag{17}$$

$$If\left[E \leq \frac{1}{2}C\right] Then\ x_i^{PIN} = CEA\left[\frac{1}{2}c_i, E\right] \tag{18}$$

$$\sum\nolimits_{i=1}^{N} x_i^{PIN} = E \tag{19}$$

In Eqs. (17) and (18), C represents the total claim of the stakeholders, x_i^{PIN} denotes the asssigned quota of stakeholder i by this method, and c_i is the claim of stakeholder i. Eq. (19) ensures that the total distributed quotas sum up to the total resource (E).

2.2.8 Groundwater-based (GW)

This method is a modified version of the CEL method, taking into account two main factors: the rate of groundwater usage (G_i) and the rate of claims (c_i). According to this method, as described by Eq. (21), the total deficit (the difference between total claims and total assets based on Eq. (20)) is allocated among stakeholders. Specifically, those who use more groundwater will see a greater difference between their claims and the quota they receive. This means that stakeholders who consume more groundwater will receive a smaller portion of the available surface water. Finally, the allocated quota for each stakeholder is determined based on Eq. (22) [26].

$$D = C - E \tag{20}$$

$$d_i = \left[\left(\frac{\left(\frac{c_i}{\sum_{i=1}^{N} c_i} \right) + \left(\frac{G_i}{\sum_{i=1}^{N} G_i} \right) + 1}{n+2} \right) \times D \right] \tag{21}$$

$$x_i^{GW} = (c_i - d_i); 0 \le x_i \le c_i \tag{22}$$

In Eq. (20), D represents the total deficit, C is the total claim of stakeholders, and E is the total amount of resources. In Eq. (21), d_i represents the deficit qouta for stakeholder i, c_i is the claim of stakeholder i, G_i denotes the current groundwater usage of stakeholder i, and N is the number of stakeholders. In Eq. (22), x_i^{GW} denotes the assigned quota for stakeholder i from total resource, and it must be positive and cannot exceed the individual claim of the stakeholder it is assigned to.

3. Application

Water scarcity and the equitable allocation of shared water resources are among the most critical challenges facing the global community today. The application of bankruptcy methods to shared water resources has seen varying degrees of success across different continents, reflecting the diverse geopolitical, social, and environmental contexts in which these methods are implemented. Case studies from around the world demonstrate how bankruptcy methods can lead to more transparent and mutually acceptable outcomes, reducing tensions and fostering long-term collaboration between water beneficiaries. By analyzing these case studies, this section aims to underscore the potential of bankruptcy methods as a tool for achieving equitable water distribution on a global scale.

In most of the research mentioned in this section, a process involving several steps is employed to determine the quotas of stakeholders sharing a water resource that has undergone bankruptcy. As illustrated in **Figure 3**, this process can be delineated into four steps:

1. Collection of primary information: Essential data is gathered based on bankruptcy method parameters to calculate quotas for stakeholders;

2. Selection of a bankruptcy method: Depending on the context, one of the bankruptcy methods is chosen. These methods fall into three categories: resource-based methods, deficit-based methods, and combined resource-deficit-based methods;

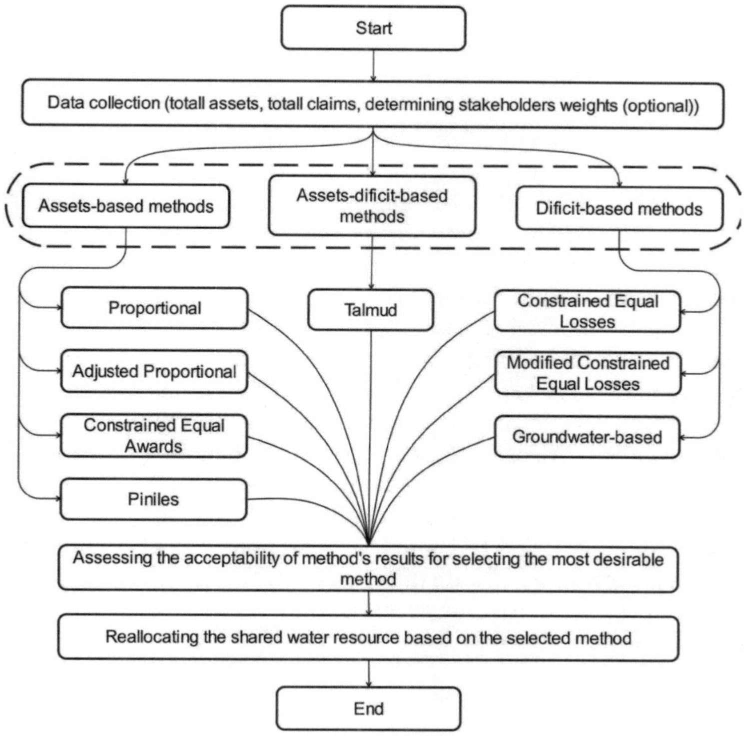

Figure 3.
Bankruptcy process involving several steps for determination of the quotas of stakeholders sharing a water resource.

3. Evaluation of results: After applying the selected bankruptcy method, the results are evaluated using various methods to measure stakeholder satisfaction with the allocated quotas;

4. Allocation of quotas: Finally, the calculated quotas are allocated to the stakeholders according to the outcomes derived from the bankruptcy method.

While this process can be adapted to suit the specific conditions of the issue at hand or the preferences of the researchers, its general structure serves as a useful framework for identifying and executing the steps of quota allocation for stakeholders in shared water resources.

The following section reviews studies that have applied these bankruptcy methods, showcasing how this framework has been utilized across different regions and scenarios to manage water conflicts and promote equitable distribution.

3.1 In Asia

3.1.1 The Yangtze river (Hubei Province, China)

The context of the water resource conflict stems from increasing water scarcity driven by global climate change, rapid industrialization, and urbanization. The situation in Ezhou City, China, exemplifies this conflict, where the projected water demand for 2030 is expected to reach 1196.85 million cubic meters annually (MCM/Y), while the planned water supply can only meet 1080.02 MCM/Y, resulting in a deficit of 116.83 MCM/Y.

The methodology section of this study outlines the different bankruptcy methods applied to the water distribution problem in Ezhou City, China. These methods include the PRO, AP, CEA, CEL, TAL, and a proposed allocation method, the adjust minimal overlap (AMO). These methods are used to calculate water allocation under different scenarios, taking into account the varying demands and priorities of different regions and sectors within the city. These rules are tested for feasibility using the "core" solution from cooperative game theory (CGT) and security restrictions tailored to the economic structure and water deficit tolerance of different regions.

The application of these bankruptcy methods to Ezhou City's water resource allocation in 2030 produced varying outcomes. Under the PRO rule, each region experienced a uniform deficit rate of 9.8%, with Gedian Development Zone (Gedian DZ) receiving the most water due to its large industrial demand. The CEA rule, however, resulted in Gedian DZ bearing the entire deficit with a rate of 23.7%, while other regions faced no shortages. The TAL and CEL rules produced identical outcomes, with Gedian DZ facing a lower deficit rate of 4.7%, while the three counties (Huarong, Echeng, and Liangzihu) suffered higher deficit rates of 18.5, 10.6, and 16.5%, respectively. The AMO rule, proposed in this study, distributed the water deficit among all stakeholders based on their claims, resulting in a deficit rate of 15.8% for Gedian DZ, 5.9% for the urban area, and rates ranging from 4.7 to 6.1% for the three counties. This method was found to balance the water distribution more equitably, sharing the burden of scarcity while considering both the size of claims and the regions' economic structures.

3.1.2 The shared aquifer (Razavi Khorasan Province, Iran)

The study focuses on the increasing conflicts over shared groundwater resources due to potable, agricultural, and industrial activities in regions such as Neyshabour, Sabzevar, and Ataiyeh in Iran [27]. This situation necessitates an approach like bankruptcy theory that not only considers the demands of all stakeholders but also ensures the equitable distribution of water resources. The study employs bankruptcy theory as the basis for reallocating shared groundwater resources by utilizing four weighted bankruptcy methods: WPRO, WCEA, WCEL, and WMCEL. To tailor these methods to the specific needs of the study area, the researchers incorporated the analytical hierarchy process (AHP) to assign weights to different stakeholders based on sustainable development criteria relevant to agriculture, drinking water, and industry.

The application of the weighted bankruptcy methods revealed that the weighted constrained equal awards (WCEA) method, based on assessing the acceptability of results, provided the most balanced and fair allocation of water resources among the stakeholders. According to the results, the Neyshabour, Sabzevar, and Ataiyeh plains were allocated 39, 68, and 45% of their total water demands, respectively.

3.1.3 The Gorgan river basin (Golestan Province, Iran)

The study focuses on maximizing agricultural profits amidst water conflicts in the Gorgan River Basin, located in Golestan Province, Iran, specifically in Minudasht, Azadshahr, and Gonbade-kavus [25]. The methodology involved developing an aquifer simulation model, calibrated with observed hydraulic head data, and creating a multilayer perceptron artificial neural network (MLP-ANN) meta-model for predicting hydraulic head. A multi-objective optimization model using the non-dominated sorting genetic algorithm (NSGA-II) was then employed to determine optimal crop cultivation patterns, balancing agricultural profits and water table preservation. Five bankruptcy methods (CEL, CEA, PRO, TAL, and AP) were used to allocate water among the stakeholders. A noncooperative 3-player game model helped identify the best management scenario, which proposed a 20% reduction in groundwater extraction and a 10% decrease in cultivation area. This scenario achieved higher profits with less water table drawdown compared to others.

3.1.4 The Indus river (among four provinces, Pakistan)

The study examines equitable water distribution among Pakistan's provinces (Punjab, Sindh, Baluchistan, and Khyber Pakhtunkhwa) with a focus on the Indus River [26]. It compares five bankruptcy methods (CEL, CEA, PRO, TAL, and PIN) and the Shapley value to address water allocation challenges. Additionally, two new methods are proposed: the "groundwater-based method", which considers groundwater use, and the "proposed method", which accounts for land salinity and provincial gross domestic product (GDP). The study finds that the CEA method favors provinces with smaller claims, CEL benefits those with larger claims, and the PRO method offers a balanced approach. Punjab receives the most water, followed by Sindh, Baluchistan, and KPK, with the groundwater-based method identified as the most suitable for the Indus River Basin.

3.1.5 The Tigris-Euphrates river basin (Turkey, Syria, and Iraq, Southwest Asia)

The study focuses on the Tigris-Euphrates River, a vital water resources shared among Turkey, Syria, and Iraq (**Figure 4**) [23].

The study introduces a novel weighted bankruptcy solution method to address water allocation challenges in the Tigris-Euphrates River Basin. This method, based on bankruptcy theory, allocates limited water resources among stakeholders (Turkey, Syria, and Iraq) by considering their contributions and relative claims. Two scenarios were analyzed: one with equal relative weights for all states and another with different weights (Turkey: 0.15, Syria: 0.3, Iraq: 0.55). For equal weights, various bankruptcy methods (PRO, CEA, and CEL) were applied, with Iraq receiving the highest allocation, especially under the CEL method. The proposed method of modified weighted constrained equal losses (MWCEL) allocated 5330 MCM/Y to Turkey, and 43,340 MCM/Y to Iraq, emphasizing Iraq's larger contribution. For different weights, the weighted methods showed significant changes: Turkey's allocation decreased under WPRO, while Iraq's increased to 45,000 MCM/Y. The MWCEL method allocated 4397 MCM/Y to Turkey, and 44,273 MCM/Y to Iraq, demonstrating the impact of relative weights on water distribution.

3.2 In Europe

3.2.1 The Tirso-Flumendosa-Campidano water system (southern Sardinia, Italy)

The study focuses on water resource allocation under critical scarcity conditions, particularly in Mediterranean regions [28]. The methodology is applied to a complex

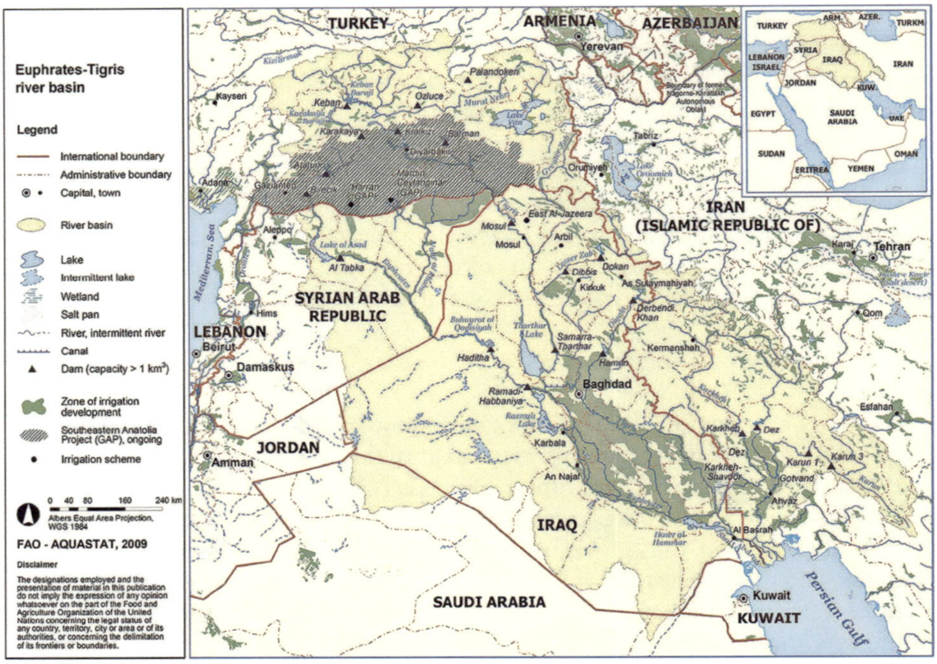

Figure 4.
The Tigris-Euphrates River Basin, the shared water resources among Turkey, Syria, and Iraq [23].

water supply system, specifically the Tirso-Flumendosa-Campidano water system in southern Sardinia, Italy. The study examines how water can be equitably allocated among beneficiaries during periods of drought or limited water availability. To address this complex problem, the study employs bankruptcy methods for water resource allocation in conditions of scarcity. Five bankruptcy methods are utilized: PRO, CEA, CEL, TAL, and AP. The methodology is applied in two phases: First, a simplified water system is analyzed, followed by a more complex case study involving the Tirso-Flumendosa-Campidano water system in southern Sardinia. Throughout the study, the results obtained from the bankruptcy methods with those derived from classical optimization and simulation modeling approaches. By verifying the application of bankruptcy methods, the study demonstrates their potential to enhance decision-making in water resource management, particularly in critical scarcity scenarios. The study finds that the TAL and AP methods provide better performance in scenarios of high-water scarcity, making them valuable references for water managers in decision-making processes.

3.3 In Africa

3.3.1 The Nile river basin (North Africa)

The study focuses on the Nile River Basin, one of the most contested shared river basins globally due to its significant socioeconomic and environmental importance [29]. The Nile River spans across eleven riparian countries in Africa (**Figure 5**), but the study focuses on eight of them, excluding South Sudan, the Democratic Republic of Congo, and Eritrea due to a lack of data. The demand for the

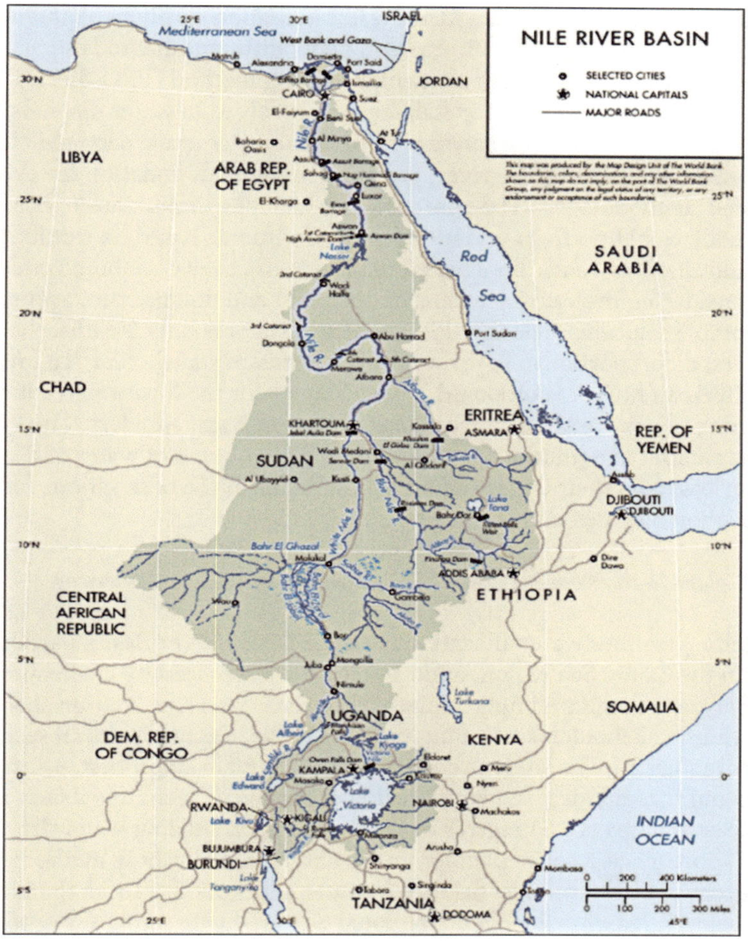

Figure 5.
The Nile River Basin, the shared water resources among the eleven riparian countries in Africa [30].

Nile's water has been increasing rapidly, driven by population growth and economic development in the basin countries. The study employs bankruptcy methods to distribute the available water in the Nile River Basin under projected scenarios of water scarcity. Four bankruptcy methods applied: PRO, CEL, CEA, and AP. The study compares the results from these bankruptcy methods with allocations using the Shapley value and nucleolus methods to assess fairness and stability in water distribution among the stakeholders.

3.4 In North America

3.4.1 The Missouri river basin (United States of America)

The study focuses on the Missouri River Basin, a critical area for understanding the complexities of water resource allocation [31]. This region has a long history of conflicts over water rights, influenced by various stakeholders, including agricultural beneficiaries, municipalities, and ecological interests. The methodology section details the analytical framework used to address water allocation conflicts. This study utilizes game theory as a foundational tool, exploring how different methods such as PRO, proportional method based on the sequential sharing method (PRO-SSR), CEA, CEL, and MCEA can affect outcomes for stakeholders involved in water disputes. The research involves quantitative analyses of historical water usage data and claims from different stakeholders along the river, providing a robust foundation for evaluating the proposed methodologies. The approach also includes simulations to predict how these methods would perform under different conditions. Based on results, the CEA and CEL allocation methods are deemed inadequate as they distribute water equally without considering the varying claims of upstream and downstream agents, leading to unacceptable solutions. The SSR-PRO, which is suitable only for linearly ordered stakeholders, disproportionately favors the downstream stakeholder. To improve upon the CEA, an MCEA is proposed, which adjusts for the lowest percentage claim rate, resulting in a more satisfactory allocation for both stakeholders. The PRO also emerges as an effective solution, as it allocates a higher quota of water to the upstream stakeholder based on their larger claim, making it one of the best options for resolving water allocation disputes in this context.

3.4.2 The Colorado river basin (California state, United States of America)

The study examines water allocation in the Colorado River Basin, particularly focusing on the Salton Sea region, amid significant water scarcity challenges [32]. It aims to analyze the welfare implications of different water distribution schemes among various stakeholders, including urban beneficiaries, agricultural sectors, and recreational interests. The study is conducted in two main steps: first, estimating water demand through demand curves using econometric techniques such as the ordinary least squares (OLS) regression, and second, optimizing water distribution using two approaches: a social planner model and two bankruptcy methods (PRO and CEA). The social planner model prioritizes water allocation based on the highest marginal benefit, maximizing overall regional welfare by directing water to sectors that generate the most economic and social value. In contrast, the bankruptcy methods, particularly the PRO method, lead to lower welfare outcomes because they do not account for the varying values of water among different beneficiaries.

4. Conclusion

This study provides an analysis of water resource allocation conflicts, focusing on shared contexts where water scarcity intensifies tensions among stakeholders. The research highlights the importance of adopting game theory, particularly bankruptcy methods, as a more effective solution than traditional optimization techniques. One primary insight from the study is the inadequacy of conventional optimization approaches in addressing the complex nature of water conflicts. This is especially true in shared water resources, where the differing claims and priorities of various countries or regions lead to significant conflicts. The findings support the use of game theory, particularly bankruptcy theory, as a more adaptable approach to the allocation of shared water resources.

Bankruptcy methods are specifically designed to address scenarios where available resources are insufficient to meet all claims. The study examines several bankruptcy methods, including proportional (PRO), adjusted proportional (AP), constrained equal awards (CEA), constrained equal losses (CEL), Talmud (TAL), Piniles (PIN), modified constrained equal losses (MCEL), and groundwater-based (GW) methods. Each method offers a distinct approach to balancing stakeholder claims, considering factors such as total available resources, the size of the claims, and the relative importance of the stakeholders.

Through detailed case studies from various regions, including Asia, Europe, Africa, and North America, the study demonstrates the practical application of these methods. In each case, bankruptcy methods have led to more mutually acceptable outcomes, reducing tensions and supporting long-term cooperation among stakeholders. For instance, in regions such as the Middle East and North Africa, these methods have successfully mediated conflicts. Similarly, in the Indus River Basin in Pakistan and the Tigris-Euphrates River Basin in Southwest Asia, bankruptcy methods have facilitated more equitable water distribution. In North America, the application of bankruptcy theory methods has effectively resolved water resources management disputes among stakeholders, as seen in the cases of the Colorado and Missouri rivers.

The study concludes that by adopting game theory and bankruptcy methods, stakeholders and policymakers can develop more effective and equitable water allocation strategies that not only address the immediate needs of stakeholders but also promote long-term sustainability and cooperation. By providing a structured approach to resource allocation that accounts for the competing interests of multiple stakeholders, bankruptcy methods can help ensure that water resources are managed in a fair and sustainable manner. The study also calls for further research and refinement of these methods, as well as their application to other water resource management challenges, including both surface and groundwater shared resources, to address the evolving needs of global water security.

Author details

Reza Javidi Sabbaghian* and Mohammad Ali Tolouei Virani
Department of Civil Engineering, Hakim Sabzevari University, Sabzevar, Iran

*Address all correspondence to: rezajs.civil.eng@gmail.com; r.javidi.s@hsu.ac.ir

References

[1] Mishra RK. Fresh water availability and it's global challenge. Journal of Marine Science: Research. 2023;**2**(1): 01-03

[2] Sabale R, Venkatesh B, Jose M. Sustainable water resource management through conjunctive use of groundwater and surface water: A review. Innovative Infrastructure Solutions. 2023;**8**(1): 1-12

[3] Sabbaghian RJ, Nejadhashemi AP. Developing a risk-based consensus-based decision-support system model for selection of the desirable urban water strategy: Kashafroud watershed study. Water (Switzerland). 2020;**12**(5):1305-1338

[4] UN-Water. Summary Progress Update 2021: SDG 6 — Water and Sanitation for all. Geneva, Switzerland: UN-Water Integrated Monitoring Initiative [Internet]; 2021. pp. 1-58. Available from: https://www.unwater.org/new-data-on-global-progress-towards-ensuring-water-and-sanitation-for-all-by-2030/

[5] MacAlister C, Baggio G, Perera D, Qadir M, Taing L, Smakhtin V. Global Water Security 2023 Assessment. Hamilton, Canada: Institute for Water, Environment & Health [Internet]; 2023. Available from: http://inweh.unu.edu/publications/

[6] Tayia A. Transboundary water conflict resolution mechanisms: Substitutes or complements. Water (Switzerland). 2019;**11**(7):1337-1376

[7] Scanlon BR, Fakhreddine S, Rateb A, de Graaf I, Famiglietti J, Gleeson T, et al. Global water resources and the role of groundwater in a resilient water future. Nature Reviews Earth and Environment. 2023;**4**(2):87-101

[8] UNECE. Policy Guidance Note on the Benefits of Shared Water Cooperation: Identification, Assessment and Communication. New York, Geneva: UNECE; 2015. 51 pp. Available from: https://www.unece.org/fileadmin/DAM/env/water/publications/WAT_Benefits_of_Shared_Cooperation/ECE_MP.WAT_47_PolicyGuidanceNote_Benefits Cooperation_1522750_E_pdf_web.pdf

[9] Simonovic SP. Managing Water Resources: Methods and Tools for a Systems Approach [Internet]. 1st ed. London: UNESCO; 2008. 680 p. Available from: https://www.taylorfrancis.com/books/mono/10.4324/9781849771917/managing-water-resources-slobodan-simonovic

[10] Mirzaei-Nodoushan F, Bozorg-Haddad O, Loáiciga HA. Evaluation of cooperative and non-cooperative game theoretic approaches for water allocation of shared rivers. Scientific Reports [Internet]. 2022;**12**(1):1-11. DOI: 10.1038/s41598-022-07971-1

[11] Akbari-Alashti H, Haddad OB, Fallah-Mehdipour E, Mariño MA. Multi-reservoir real-time operation rules: A new genetic programming approach. Proceedings of the Institution of Civil Engineers: Water Management. 2014; **167**(10):561-576

[12] Tayfur G. Modern optimization methods in water resources planning, engineering and management. Water Resource Management. 2017;**31**(10): 3205-3233

[13] Madani K. Game theory and water resources. Journal of Hydrology [Internet]. 2010;**381**(3–4):225-238. DOI: 10.1016/j.jhydrol.2009.11.045

[14] Harsanyi JCSR. A General Theory of Equilibrium Selection in Games. 1st ed. Cambridge: MIT Press Books; 1988

[15] Myerson R. Analysis of conflict. In: Game Theory. Cambridge: Harvard Univetsity Press; 1991

[16] Swathi N. Water resource conflicts: A game theory approach. Malaya Journal of Matematik. 2020;**S**(2):1416-1420

[17] Bogardi I, Szidarovsky F. Application of game theory in water management. Applied Mathematical Modelling. 1976; **1**(1):16-20

[18] Nash JF. Equilibrium points in n-person games. Proceedings of the National Academy of Sciences. 1950; **36**(1):9-48

[19] Dinar A, Nigatu GS. Distributional considerations of international water resources under externality: The case of Ethiopia, Sudan and Egypt on the Blue Nile. Water Resources and Economics [Internet]. 2013;**2–3**:1-16. DOI: 10.1016/j.wre.2013.07.001

[20] Zheng Y, Liu Z, Corporation G, Zhang S, Liu P, Corporation G. Water allocation management under scarcity: A bankruptcy approach. Water Resource Management [Internet]. 2022;**36**(9): 2891-2912. Available from: https://link.springer.com/article/10.1007/s11269-022-03098-0#citeas

[21] Shahraki AS, Singh VP, Bazrafshan O. Developing a bankruptcy theory to resolve stakeholders' conflict over optimal water allocation: The case of Hirmand catchment. Water (Switzerland). 2024;**16**(9):1303-1328

[22] Aumann RHS. Handbook of game theory: With economic applications. Technical Report. 2016;**103**(421): 1559-1560

[23] Mianabadi H, Mostert E, Pande S, van de Giesen N. Weighted bankruptcy rules and shared water resources allocation. Water Resources Management. 2015;**29**(7):2303-2321

[24] Zheng Y, Sang X, Liu Z, Zhang S, Liu P. Water allocation management under scarcity: A bankruptcy. Water Resource Management [Internet]. 2022; **36**:2891-2912. DOI: 10.1007/s11269-022-03098-0

[25] Yazdian M, Rakhshandehroo G, Nikoo MR, Mooselu MG, Gandomi AH, Honar T. Groundwater sustainability: Developing a non-cooperative optimal management scenario in shared groundwater resources under water bankruptcy conditions. Journal of Environmental Management [Internet]. 2021;**292**(March):112807. DOI: 10.1016/j.jenvman.2021.112807

[26] Janjua S, Hassan I. Use of bankruptcy methods for resolving interprovincial water conflicts over shared river: Case study of Indus River in Pakistan. River Research and Applications [Internet]. 2020;**36**(7): 1334-1344. Available from: https://onlinelibrary.wiley.com/doi/abs/10.1002/rra.3621

[27] Bahrami JB, Javidi SR, Tolouei VMA, Roghani BFM. Development of a model for reallocation of shared aquifer resources among stakeholders based on the bankruptcy approach; case study of Neyshabour-Sabzevar-Ataiyeh. Iran-Water Resources Research [Internet]. 2023;**19**:39-56. Available from: https://www.iwrr.ir/article_158798.html

[28] Sechi GM, Zucca R. Water resource allocation in critical scarcity conditions: A bankruptcy game approach. Water Resources Management. 2014;**29**: 541-555. Available from: http://wej.miau.ac.ir/article_4913.html?lang=en

[29] Degefu DM, He W. Allocating water under bankruptcy scenario. Water Resource Management [Internet]. 2016; **30**(11):3949-3964. DOI: 10.1007/s11269-016-1403-x

[30] Mulugeta D, Weijun D. Water bankruptcy in the mighty Nile river basin. Sustainable Water Resources Management. 2016;**2**(1):29-37

[31] Wickramage HM, Roberts DC, Hearne RR. Water allocation using the bankruptcy model: A case study of the Missouri river. Water. 2020;**12**(3):619

[32] Rightnar J, Dinar A. The welfare implications of bankruptcy allocation of the Colorado river water: The case of the Salton Sea region. Water Resources Management. 2020;**34**(8):2353-2370

Section 4

Integrative and Interdisciplinary Approaches

Chapter 7

Integrating Scientific Instruments and Intelligence Frameworks for Digital Enterprises

Eyup Akcetin

Abstract

In the rapidly evolving business landscape, integrating methodologies from diverse fields is essential for optimizing decision-making and achieving strategic goals. This chapter demonstrates how systems thinking, game theory, business intelligence, decision intelligence, strategic intelligence, and swarm intelligence work synergistically to enhance business performance. Systems thinking analyzes organizational relationships and feedback loops, while game theory models strategic interactions and competitive behaviors. Business intelligence derives actionable insights from data, and decision intelligence uses advanced analytics to improve predictive capabilities. Strategic intelligence helps businesses plan long-term strategies, and swarm intelligence offers decentralized solutions for optimization and problem-solving. The integration of scientific tools such as thermometers and barometers into business contexts further enhances performance measurement. For example, employee satisfaction can be measured as a 'temperature' and market pressures as a 'barometer.' These interdisciplinary approaches, enriched with big data analytics, provide businesses with deep insights, improve decision-making, and foster sustainable innovation. In the real world, companies such as Apple leverage these combined frameworks to gain a competitive edge in their product development and market strategies. By adopting these tools, businesses can achieve faster decision-making, strategic foresight, and long-term success.

Keywords: system thinking, game theory, business intelligence, decision intelligence, strategic intelligence, swarm intelligence

1. Introduction

Adapting measurement instruments from various scientific fields into business sciences enables companies to develop more effective and efficient management strategies in complex and dynamic business environments. This adaptation not only optimizes business processes but also complements strategic approaches like System Thinking and Game Theory, creating a holistic management strategy.

Measurement instruments inspired by fields such as physics, chemistry, engineering, and environmental sciences can be used to evaluate and optimize business functions' performance. For example, thermometers can measure employee satisfaction and motivation, helping businesses make more informed decisions in human resource management. Combining these tools with other interdisciplinary methods offers businesses a complete perspective on internal and external challenges. Similarly, barometers can analyze market pressures and economic conditions, providing critical data for strategic planning and gaining competitive advantage. Compasses and altimeters can determine a business's strategic direction and the height of its success, guiding long-term goal achievement.

The principles of these measurement instruments, when adapted to business sciences, contribute to increased efficiency, innovation capacity, and competitive strength, thereby ensuring sustainable growth and success. Big data analytics further enhances the accurate and comprehensive use of these measurement and evaluation tools by providing businesses with deep insights. Through big data, companies can collect and analyze real-time information from extensive databases, improving the accuracy and reliability of data obtained from these measurement tools. Thus, by adopting an interdisciplinary approach and integrating big data technologies, enriching measurement and evaluation processes allows businesses to adapt to changing market conditions and gain strategic advantages. Big data expands companies' measurement and evaluation capacities, enabling these tools to be used more effectively and allowing for deeper and more strategic decision-making in business management.

Systems thinking is a critical skill for analyzing complex issues, particularly in the context of organizational management. By understanding the interconnections and relationships among components within a system, it enables a comprehensive analysis of system outcomes and behaviors [1]. This approach is essential for identifying dynamic relationships, organizing components effectively, and comprehending the cyclic nature of systems [2]. In the context of health, systems thinking is valuable for analyzing interactions between components to explain system outcomes and behaviors [3]. This approach aids in reconceptualizing health problems, goals, and potential policy solutions [4]. In education, particularly in chemistry, systems thinking is essential for developing effective strategies and understanding complex systems [5]. It involves visualizing interconnections, examining changing behaviors over time, and exploring how system-level phenomena emerge from interactions between parts [6]. Furthermore, systems thinking is recommended in healthcare to enhance quality improvement and patient safety efforts [7]. In organizational management, systems thinking facilitates improved decision-making, innovation, and sustainability by providing valuable theories, models, and approaches to navigate complexity [8, 9]. The application of systems thinking extends beyond education and healthcare. Moreover, systems thinking provides valuable theories, models, and approaches for comprehending complexity in mental health, partnerships, and general systems. In conclusion, systems thinking is a versatile approach that transcends disciplines, offering a holistic perspective to analyze and address intricate problems. By emphasizing relationships, dynamics, and emergent phenomena within systems, individuals and organizations can navigate complexity more effectively and make informed decisions. Therefore, systems thinking complements business measurement tools inspired by other sciences, such as physics and engineering, enhancing strategic decision-making and fostering long-term success. When combined with Game Theory, this framework not only addresses organizational complexities but also highlights the strategic competitive dynamics within industries.

Game theory, a field with applications in various disciplines, offers a framework for analyzing strategic interactions among rational decision-makers. It has been integrated with economics, providing insights into decision-making processes and equilibrium outcomes [10]. Moreover, game theory has found applications in computer science, particularly in the development of computational systems and information security [11]. The study of game theory enables individuals to make predictions, solve problems, and understand equilibrium strategies within strategic interactions [12]. Evolutionary game theory, a subset of game theory, explores strategic behavior in evolving populations. These evolving strategies, when informed by Systems Thinking, allow organizations to adapt and anticipate future changes in competitive landscapes. It considers the impact of frequency-dependent selection and the underlying structure of networks on the outcomes of evolutionary games. This branch of game theory has diverse applications, ranging from ecology to economics [13]. Additionally, evolutionary graph theory has been reviewed in the context of game theory, highlighting its relevance in understanding population dynamics and strategic interactions [14]. Systems thinking, when applied to game design, can enhance the development of games that foster holistic thinking and problem-solving skills. By incorporating elements of systems thinking into game design, limitations of traditional research methodologies can be overcome, leading to more transferable findings and improved outcomes [15]. Furthermore, game-based learning from a complexity thinking perspective emphasizes the alignment of game characteristics with those of complex learning systems [16]. In conclusion, game theory serves as a powerful tool for analyzing strategic interactions and decision-making processes across various disciplines. Its integration with systems thinking and evolutionary dynamics enriches the understanding of complex systems and strategic behaviors, offering valuable insights into population dynamics, computational systems, and educational methodologies. Therefore, the integration of game theory with measurement tools inspired by various sciences, such as physics and engineering, complements systems thinking, enhancing strategic decision-making and fostering long-term success across multiple disciplines. By integrating the complexity of Systems Thinking with the predictive power of Game Theory, businesses can develop more sustainable, long-term strategies.

Business intelligence, an essential aspect of modern organizations, utilizes various tools and methodologies to improve decision-making processes. Game theory, a mathematical framework that analyzes strategic interactions, plays a significant role in shaping business intelligence practices. By modeling rationality in marketing decision-making, game theory enables firms to predict competitive behavior accurately and make informed decisions swiftly [17]. Moreover, game theory has been applied to family business succession, providing insights into the complexities of succession processes and events [18]. In the realm of independent game publishing, business intelligence technologies have been crucial in game development for design and optimization purposes [19]. Analyzing game telemetry data using pattern recognition and statistical analysis offers valuable insights for game development, highlighting the role of business intelligence tools in this context [20]. These same principles can be applied in business settings, where BI helps monitor and optimize organizational processes in real time. Additionally, game theory, particularly cooperative game theory, has been utilized for community detection in social networks, demonstrating its applicability in understanding complex interactions [21]. The integration of game theory with business intelligence extends to various domains. For instance, game theory serves as a primary tool for predicting economic behavior and designing incentive mechanisms for integrating business applications between organizations [22]. Furthermore, game theory has been instrumental in optimizing system performance, particularly in smart cities, by leveraging social ties and consumer preferences [23]. Together with Business Intelligence, these tools can provide predictive insights into how decisions ripple through both the competitive landscape and consumer behavior. The application of game theory in wireless sensor networks' security requirements and threat mitigation underscore its versatility in addressing scalability and security challenges [24]. In conclusion, the integration of game theory into business intelligence practices offers a robust framework for analyzing strategic interactions, predicting behaviors, and optimizing decision-making processes across diverse domains. By leveraging the principles of game theory, organizations can gain valuable insights, enhance strategic planning, and drive innovation in today's dynamic business landscape. Consequently, incorporating game theory into business intelligence practices, alongside measurement tools inspired by various scientific disciplines, enables organizations to navigate complex strategic interactions, optimize decision-making, and foster innovation for sustained success.

Decision intelligence, a crucial element of organizational decision-making processes, utilizes various tools and methodologies to enhance strategic choices. Business intelligence systems play a pivotal role in facilitating informed decision-making within organizations. These systems encompass processes that are based on well-informed decisions, leading to improved performance levels [25]. By providing decision-makers with timely and relevant information, business intelligence tools enable organizations to act swiftly and make optimal decisions in dynamic business environments [26]. The adoption of business intelligence systems has been increasing, with a focus on understanding the factors influencing such decisions [27]. Organizations are increasingly depending on decision support and analytics provided by business intelligence to drive strategic decision-making processes [28]. These systems are designed to offer actionable insights derived from data analysis, aiding in making informed decisions across various operational functions [29]. Moreover, the impact of critical success factors in business intelligence extends beyond tools and software, encompassing processes and methodologies crucial for enhancing decision-making and organizational outcomes [30]. The integration of business intelligence with corporate strategic management has been highlighted as a means to improve decision-making processes. By utilizing data mining and balanced scorecard approaches, organizations can enhance their strategic decision-making capabilities [31]. Real-time business intelligence systems further enable organizations to make better and timelier decisions by presenting relevant information when and where decisions need to be made. These systems support fact-based decision-making by providing pre-analyzed and documented data for informed choices [32]. In conclusion, business intelligence systems serve as a cornerstone for decision intelligence within organizations, offering valuable insights, analytics, and support for strategic decision-making processes. By leveraging the capabilities of business intelligence tools and methodologies, organizations can enhance their decision-making efficiency, drive performance improvements, and gain a competitive edge in today's dynamic business landscape. Thus, integrating decision intelligence with measurement tools inspired by various scientific fields enhances business intelligence practices, enabling organizations to make more strategic and informed decisions for sustainable success.

Strategic intelligence plays a crucial role in organizational decision-making processes, enabling firms to make well-informed and effective strategic choices. Business intelligence tools are essential for enhancing strategic decision-making within organizations by providing valuable insights derived from data analysis [29]. These tools empower decision-makers to act swiftly and make optimal decisions in dynamic business environments [26]. The adoption of business intelligence systems has been on the rise, with a focus on understanding the factors influencing decision-making processes [27]. Furthermore, integrating business intelligence with corporate strategic management has been identified as a strategy to enhance decision-making processes. Through the utilization of data mining and balanced scorecard approaches, organizations can improve their strategic decision-making capabilities [31]. Real-time business intelligence systems also enable organizations to make better and timelier decisions by presenting relevant information when and where decisions need to be made. These systems support fact-based decision-making by providing pre-analyzed and documented data for informed choices [32]. In addition to business intelligence, game theory offers valuable insights into strategic behavior in competitive business environments [33]. Game theory provides a mathematical framework to model and analyze strategic interactions between decision-makers, aiding in predicting behaviors and optimizing decision-making processes [34]. The study of strategic and extensive forms of non-cooperative game theory contributes to understanding human behavior in strategic situations based on experimental data [35]. Moreover, the role of intelligence in strategic games is explored, proposing new strategies that depend on intelligence information about opponents' actions [36]. In conclusion, the synergy between business intelligence tools and game theory enhances strategic intelligence within organizations, enabling them to make well-informed decisions, predict behaviors, and optimize strategic choices. By leveraging the capabilities of business intelligence systems and game theory frameworks, organizations can navigate complex strategic environments, drive performance improvements, and gain a competitive edge in the ever-evolving business landscape. Thus, the integration of strategic intelligence with measurement tools inspired by various scientific disciplines, coupled with business intelligence and game theory, enhances organizations' ability to make informed decisions, predict behaviors, and optimize strategic choices for long-term success.

The Interrelationship Between These Concepts: Systems Thinking provides a comprehensive framework for understanding the complexity and interdependencies within a system, while Game Theory offers tools for analyzing strategic interactions and decision-making. Business Intelligence supplies the necessary data and insights to inform decisions and understand system performance. Decision Intelligence enhances decision-making by applying advanced analytics and considering the broader system context. Strategic Intelligence leverages insights from all these approaches to formulate and execute effective long-term strategies. Swarm Intelligence contributes decentralized, adaptive solutions for optimization and decision-making by leveraging the collective behavior of simple agents. Together, these concepts create a robust framework for understanding and managing complexity, making informed decisions, and achieving strategic objectives in dynamic and competitive environments. Therefore, the integration of systems thinking, game theory, business intelligence, decision intelligence, and swarm intelligence, alongside measurement tools inspired by various scientific disciplines, provides a comprehensive and adaptive framework for strategic decision-making and achieving long-term success in dynamic environments.

2. Principles of systemic thinking: Understanding interconnectedness and complexity

Systems Thinking: Interconnectedness and Unintended Consequences: Systems Thinking (ST) emphasizes that nothing exists in isolation; every element is part of an interconnected whole. The unexpected interactions of subsystems often lead to surprising outcomes. This interconnectedness is evident across various domains, highlighting the importance of understanding relationships and connections. Interconnectedness and Systemic Risk At the core of ST is the recognition that a system is composed of separate yet interconnected parts that interact to ensure proper functioning [37]. The recent financial crisis underscores that interconnectedness within the financial system is a significant source of systemic risk [38]. Externalities driven by cyclical and structural vulnerabilities contribute to interconnected structures among financial intermediaries and markets, potentially leading to financial contagion [39]. Understanding these relationships is crucial, emphasizing that everything is defined by its connections to other elements [40].

Ripple Effects and Unintended Consequences In systems, every action creates ripple effects, leading to unforeseen consequences. For instance, unintended consequences can arise in conservation management efforts, where species introductions or eradications affect biodiversity and ecosystem dynamics [41]. Similarly, unintended gender effects of international development initiatives highlight the need for a comprehensive understanding of their broader implications [42]. In organizational change, cultural interventions can lead to unintended consequences, impacting organizational dynamics and outcomes [43]. These examples underscore the intricate web of relationships and interactions within systems, where every action can trigger a chain of events with wide-ranging effects.

Changing the Rules for Better Outcomes Different players in the same structure produce similar results. Rather than trying to control the players, it is more effective to change the rules. In public finance and politics, the structure of governance systems, such as parliamentary versus presidential-congressional regimes, can lead to distinct outcomes in terms of redistribution, public goods provision, and rents to politicians [44]. This highlights how altering institutional rules can influence behavior and outcomes. Similarly, in sports, adjusting game rules or formats can optimize player performance and outcomes, showcasing the importance of rule changes in influencing behavior and results [45].

Empowering Through Understanding One element can impact the whole system, and the system's boundaries are defined by the main actor. In organizational change, boundary work plays a crucial role in transforming fields [46]. Managing boundaries and practices can influence dynamics and outcomes within the organizational ecosystem. In healthcare, boundary organizing drives practice change and incorporates new actors and resources [47]. Recognizing the impact of individual elements on the entire system and leveraging these dynamics can drive innovation and collaboration.

Both/And Thinking for Innovation Adopting a "Both/And" mindset instead of "Either/Or" combines opposing solutions to provide new perspectives. Understanding both the strengths and weaknesses of a structure empowers individuals to navigate complexities effectively. Integrating diverse network architectures, for example, harnesses the strengths of strong and weak ties while mitigating associated risks [48]. This approach optimizes performance and resilience by leveraging the unique advantages of each structure.

Waste Management and Resource Recovery Understanding that there is no "away" and knowing where waste goes can turn trash into treasure. Innovative waste management strategies and resource recovery techniques, such as the thermal treatments of animal manures for phosphorus recovery [49], demonstrate how waste can be transformed into valuable resources. Embracing such approaches contributes to sustainable practices and environmental conservation.

Root Causes and System Understanding Avoiding superficial solutions and understanding root causes is crucial for effective decision-making. Establishing cause-and-effect relationships helps in understanding systems and their actors. For example, identifying enablers and causal relationships among them can assist small organizations in prioritizing key factors for innovation [50].

Anticipating Unforeseen Changes Small events can create significant impacts. Understanding the unforeseen and establishing cause-and-effect relationships is essential for navigating complex systems. Research on cascading events and their implications [51] highlights the importance of addressing root causes to enhance preparedness and response strategies.

Embracing Complexity in Decision-Making Recognizing that models and theories cannot capture all complexities is fundamental. Good decisions made on paper can fail in the field. Understanding the complexities of collaborative and networked approaches in healthcare practices [52] emphasizes the importance of navigating dynamic and interconnected environments.

Continuous Inquiry and Exploration Every answer leads to new questions, and there are no final answers. Continuous exploration and inquiry are necessary to comprehend complex phenomena, as illustrated by research on children's scientific curiosity [53].

Systems Thinking provides a powerful framework for understanding interconnectedness, anticipating unintended consequences, and making informed decisions in complex environments. By recognizing the intricate relationships within systems, individuals, and organizations can navigate complexity more effectively and drive innovation and progress.

3. Strategic decision-making through system thinking and game theory

By integrating Systems Thinking, Game Theory, Business Intelligence, Decision Intelligence, and Strategic Intelligence, we can make more effective and efficient decisions. Here's how to combine these concepts:

3.1 Start with systems thinking

Systems Thinking helps us understand how different components interact and influence each other. By using this approach, you can map out all the components of your business or organization, identify feedback loops, relationships, and dynamics within the system, and detect key issues and opportunities. Equipping business managers with systems thinking skills to understand complex systems and their behaviors is crucial. This comprehensive understanding of interactions and dynamics can be highly beneficial for effectively managing and optimizing the system [54].

3.2 Analyzing strategic interactions with game theory

Game Theory is a crucial tool for understanding competitive and cooperative environments. By using this approach, you can analyze the potential strategies of competitors and partners, create various scenarios to identify the best strategies and develop optimal strategies. Game theory examines the evolutionary dynamics of strategic interactions, exploring how lower payoff strategies are replaced by higher payoff strategies over time. This understanding provides valuable insights into decision-making processes and competitive behaviors. Thus, game theory supports the analysis of strategic interactions, the determination of optimal strategies, and the development of effective strategies [55].

3.3 Gathering data and insights with business intelligence

Business Intelligence (BI) enables data-driven decision-making and provides solutions to challenges encountered in the fields of business intelligence and performance management. The integration of BI with System Thinking and Game Theory further strengthens the strategic decision-making process by providing a data-backed understanding of complex interactions. By utilizing BI tools and performance management strategies, organizations can collect and analyze data on business performance, market trends, and customer behavior. In this process, BI tools are used to visualize the data and gain insights. This visualization process, when combined with strategic intelligence, helps decision-makers to not only track past performance but also to forecast future trends with greater accuracy. These insights allow for informed, data-driven decisions, enabling organizations to make more effective and efficient decisions [56].

3.4 Optimize decision-making with decision intelligence

Decision Intelligence uses advanced analytics and artificial intelligence to enhance decision-making processes. By adding Decision Intelligence into this framework, businesses can enhance their predictive capabilities, reducing uncertainty in decision-making processes. By employing advanced analytical models and machine learning algorithms, organizations can make accurate predictions and forecasts. Conducting scenario analyses allows for the evaluation of different possibilities and their outcomes, thereby making decision-making processes more effective and efficient. Leveraging decision intelligence tools and organizational ambidexterity enables organizations to develop strategies that optimize decision-making processes effectively [57].

3.5 Developing long-term plans with strategic intelligence

Strategic Intelligence supports long-term planning and decision-making by enhancing organizations' strategic decision-making capabilities. When combined with System Thinking, Strategic Intelligence helps businesses not only plan for the long-term but also to remain adaptable in complex, ever-changing environments. This approach involves setting long-term strategic goals, developing plans to achieve these goals, and analyzing the competitive landscape and future trends to identify strategic opportunities and threats. It also integrates insights from various approaches to create comprehensive and coherent strategies. Incorporating strategic intelligence into the planning process enables organizations to effectively achieve their long-term goals [58].

3.6 Decision-making and optimization with swarm intelligence

Swarm Intelligence has made significant advancements and offers a wide range of applications in decision-making, optimization, and problem-solving across various fields. The integration of Strategic Intelligence with Swarm Intelligence offers businesses a powerful tool for decentralized decision-making and adaptive problem-solving. For instance, a comprehensive review of UAV swarm intelligence discusses its recent advances and future trends in decision-making, path planning, control, communication, and application layers [59]. Together, Swarm Intelligence and Game Theory create a dynamic framework for navigating competitive environments and optimizing decisions through collective behavior. The beetle swarm optimization algorithm exemplifies superior intelligence and adaptability in route planning, demonstrating its effectiveness in optimizing complex planning tasks [60]. The power of swarm intelligence in solving optimization and decision-making problems is highlighted, emphasizing its efficacy in addressing complex challenges [61]. Furthermore, advancements in swarm intelligence algorithms for RFID network planning in smart factories are examined, with a focus on optimization techniques and their impact on network performance [62]. The role of swarm intelligence in decision-making processes is also underscored, illustrating how intelligent algorithms enhance decision outcomes. Collectively, these references underscore the effectiveness and broad application potential of swarm intelligence in problem-solving, optimization, and decision-making processes [63].

4. Enhancing business decision-making through the integration of advanced intelligence frameworks

Objective: To enhance business decision-making and strategic planning by integrating System Thinking, Game Theory, Business Intelligence, Decision Intelligence, Strategic Intelligence, and Swarm Intelligence.

Overview: In today's complex business environment, companies need to adopt advanced methodologies to remain competitive and achieve their strategic objectives. Integrating System Thinking, Game Theory, Business Intelligence, Decision Intelligence, Strategic Intelligence, and Swarm Intelligence provides a comprehensive approach to optimize decision-making processes, increase efficiency, and enhance strategic planning. To further illustrate the synergy of these frameworks, practical examples from different industries can show how combining these intelligence leads to innovation and strategic advantages.

Proposed strategy:

1. Implement System Thinking for Holistic Analysis:
    - *Objective:* Understand the broader context and interdependencies within the organization.
    - *Action:* Map out all components of the business, and identify feedback loops, relationships, and dynamics.
    - *Benefit:* Enables the company to anticipate potential issues and optimize operations by understanding systemic interactions.
2. Apply Game Theory for Strategic Interactions:
    - *Objective:* Analyze competitive and cooperative behaviors among market participants.

- *Action:* Use game theory models to formulate strategies that maximize outcomes while anticipating competitors' actions.
- *Benefit:* Improves the company's ability to navigate competitive environments and make strategic decisions.

3. Utilize Business Intelligence for Data-Driven Decisions:
   - *Objective:* Gain insights from business data to support decision-making.
   - *Action:* Implement BI tools to collect, integrate, analyze, and present data.
   - *Benefit:* Identifies trends, improves operational efficiency, and enhances competitive advantage through data-driven decisions.
   - *Case studies: Nature-Inspired Optimization;* use nature-inspired algorithms to reduce costs and increase efficiency in supply chain management [64]. *Swarm Intelligence Algorithms for Feature Selection;* use swarm intelligence algorithms in data analytics processes to achieve better results [65].
4. Leverage Decision Intelligence for Advanced Analytics:
   - *Objective:* Enhance decision-making capabilities using advanced analytical techniques.
   - *Action:* Employ data science, machine learning, and predictive analytics to reduce uncertainty in decision-making.
   - *Benefit:* Increases the likelihood of successful outcomes and enables quick, informed decisions.
   - *Case study: Decision Making in Brains and Social Insect Colonies;* employ optimal decision-making strategies from nature in business contexts for quick and accurate decisions [66].
5. Develop Strategic Intelligence for Long-Term Planning:
   - *Objective:* Support long-term planning and strategic decision-making.
   - *Action:* Systematically collect and analyze information on market trends, competitive dynamics, and external influences.
   - *Benefit:* Aligns strategies with long-term goals and objectives, ensuring sustained business growth.
6. Incorporate Swarm Intelligence for Collaborative Problem-Solving:
   - *Objective:* Optimize business processes through decentralized, self-organized systems.

- *Action:* Implement swarm intelligence algorithms for problem-solving and decision-making.
- *Benefit:* Enhances efficiency and effectiveness in complex environments by leveraging collective intelligence.
- *Case studies: Collective Cognition and Decision Making;* leverage the collective intelligence of employees to make better strategic decisions [67]. *Artificial Swarm Intelligence for Knowledge Pooling;* combines employee knowledge and experiences using artificial swarm intelligence for more accurate collective decisions [68].

Integrating System Thinking, Game Theory, Business Intelligence, Decision Intelligence, Strategic Intelligence, and Swarm Intelligence offers a powerful framework for businesses to enhance their decision-making processes, optimize operations, and achieve strategic objectives. By adopting these advanced methodologies, companies can gain a competitive advantage and ensure sustainable growth in a complex business environment.

4.1 Implementing system thinking for enhanced organizational performance

Objective: To enhance business performance and sustainability by integrating System Thinking principles, enabling a holistic understanding of the organization, its supply chain, and its environment.

Overview: System Thinking provides a comprehensive framework for analyzing the interrelationships and dynamics within an organization. By mapping out all components, identifying feedback loops, and understanding the broader context, businesses can make more informed decisions and improve overall efficiency. The following strategy outlines how System Thinking can be effectively integrated into business practices.

Proposed strategy:

1. Conduct a Systems Thinking Inventory:
    - *Objective:* Identify key components, feedback loops, and relationships within the organization.
    - *Action:* Utilize the methodology from Sweeney and Sterman's study on "Bathtub Dynamics" to map out circular cause-effect relations and understand the whole system [69].
    - *Benefit:* Provides a clear visualization of the organization's structure and dynamics, highlighting potential areas for improvement.
2. Measure Supply Chain Performance:
    - *Objective:* Enhance understanding of the entire supply chain and identify key interdependencies.

- *Action:* Apply Beamon's framework for measuring supply chain performance to map out the supply chain, and identify critical interdependencies, and feedback loops [70].
- *Benefit:* Improves supply chain efficiency by identifying and addressing bottlenecks and inefficiencies.

3. Apply Systems Thinking to Specific Initiatives:
   - *Objective:* Address specific organizational challenges using systems thinking principles.
   - *Action:* Utilize the approach from Owen et al.'s research on obesity prevention to tackle issues such as operational inefficiencies or customer satisfaction. Focus on understanding feedback loops and non-linear structures related to the challenge [71].
   - *Benefit:* Provides targeted solutions that consider the broader system context, leading to more sustainable outcomes.
4. Integrate Systems Thinking in Innovation Project Management:
   - *Objective:* Enhance problem-solving and decision-making in innovation projects.
   - *Action:* Implement the insights from Kapsali's study to apply systems thinking in managing innovation projects. Focus on a holistic approach to problem-solving and decision-making [72].
   - *Benefit:* Increases the success rate of innovation projects by considering all aspects of the project and their interdependencies.
5. Implement Systems Thinking in Supply Chain Management:
   - *Objective:* Improve supply chain performance through the application of systems thinking.
   - *Action:* Follow the recommendations from Wilden and Sadler's systematic literature review to incorporate systems thinking practices into supply chain management [73].
   - *Benefit:* Enhances supply chain performance by adopting a systemic approach to problem-solving and continuous improvement.

Integrating System Thinking into business practices offers a powerful approach to understanding and improving organizational performance. By mapping out components, identifying feedback loops, and considering the broader context, businesses can make more informed decisions, enhance efficiency, and drive sustainable practices. The proposed strategy leverages key insights from relevant studies to create a comprehensive framework for implementing System Thinking in a business context. This comprehensive integration of scientific tools and intelligence provides businesses with the necessary strategic advantage to succeed in a highly competitive and complex environment.

4.2 Implementing game theory for enhanced competitive strategy and decision-making

Objective: To enhance competitive strategy and decision-making by integrating Game Theory principles, enabling a deeper understanding of strategic interactions and optimal decision-making in business environments.

Overview: Game Theory analyzes strategic interactions and decision-making processes, providing a robust framework for evaluating potential strategies of competitors and partners. By developing various scenarios and determining optimal strategies under different conditions, businesses can improve their competitive strategy and negotiation processes. The following strategy outlines how Game Theory can be effectively integrated into business practices.

Proposed strategy:

1. Analyze Competitor Strategies and Market Scenarios:

    - *Objective:* Gain insights into competitors' potential strategies and market dynamics.

    - *Action:* Utilize the insights from Meyer's study on "Quantum Strategies" to explore advanced competitive scenarios and strategic interactions [74].

    - *Benefit:* Helps businesses anticipate competitor moves and develop robust strategies to stay ahead in the market.

2. Develop Competitive Strategies and Contingency Plans:

    - *Objective:* Create effective competitive strategies and prepare for various market conditions.

    - *Action:* Apply the methodologies from Ho et al.'s research on game theory in defense applications to develop comprehensive competitive strategies and contingency plans [75].

    - *Benefit:* Enhances the business's ability to adapt to changing market conditions and mitigate risks.

3. Optimize Marketing-Mix Strategies:

    - *Objective:* Plan and implement optimal marketing-mix strategies using game theory approaches.

    - *Action:* Implement the approach from Abedian et al.'s study to select the best marketing-mix strategies based on game theory analysis [76].

- *Benefit:* Improves marketing effectiveness by considering competitors' actions and market responses.

4. Apply Game Theory in Economic and Industrial Contexts:

   - *Objective:* Utilize game theory to understand economic and industrial dynamics.

   - *Action:* Leverage insights from Nie et al.'s research on the application of game theory in economics to inform business strategy and decision-making processes [77].

   - *Benefit:* Provides a theoretical foundation for strategic planning and enhances the understanding of market structures and behaviors.

5. Determine Marketing Strategies Based on Human-Computer Interaction:

   - *Objective:* Develop marketing strategies using game theory and human-computer interaction insights.

   - *Action:* Follow the approach from Christanto's study to determine marketing strategies for digital platforms based on usability and game theory analysis [78].

   - *Benefit*: Enhances the effectiveness of digital marketing strategies by incorporating user behavior and competitive dynamics.

Integrating Game Theory into business practices offers a powerful approach to understanding and improving competitive strategy and decision-making. By analyzing competitor strategies, developing various scenarios, and determining optimal strategies, businesses can make more informed decisions and gain a competitive advantage. The proposed strategy leverages key insights from relevant studies to create a comprehensive framework for implementing Game Theory in a business context.

4.3 Implementing business intelligence for enhanced decision-making and performance monitoring

Objective: To enhance decision-making and performance monitoring by integrating Business Intelligence (BI) principles, enabling the collection, analysis, and visualization of data on business performance, market trends, and customer behavior.

Overview: Business Intelligence provides the data and insights necessary for informed decision-making. By collecting and analyzing data, visualizing it through BI tools, and extracting actionable insights, businesses can support decision-making processes with accurate and real-time information. The following strategy outlines how BI can be effectively integrated into business practices.

Proposed strategy:

1. Collect and Analyze Sales Data, Market Trends, and Customer Preferences:

   - *Objective:* Gather comprehensive data to support business decisions.

- o *Action:* Utilize the insights from Feng et al.'s study on integrated thinking in integrated reporting to establish a robust data collection and analysis framework [79].

- o *Benefit:* Provides a detailed understanding of business performance, market trends, and customer behavior, enabling data-driven decisions.

2. Implement BI Dashboards for Real-Time Performance Monitoring:

   - o *Objective:* Monitor real-time performance metrics to make timely and informed decisions.

   - o *Action:* Apply methodologies from Guthrie et al.'s research on integrated reporting and integrated thinking to develop BI dashboards that track key performance indicators (KPIs) [80].

   - o *Benefit:* Enhances the ability to monitor and respond to performance metrics in real-time, improving operational efficiency.

3. Use Integrated Reporting for Comprehensive Insights:

   - o *Objective:* Provide a holistic view of the organization's performance and strategy.

   - o *Action:* Implement integrated reporting practices based on insights from Al-Htaybat and Alberti-Alhtaybat's study to combine financial and non-financial data for comprehensive insights [81].

   - o *Benefit:* Supports informed decision-making by providing a complete picture of the organization's performance, risks, and opportunities.

Integrating Business Intelligence into business practices offers a powerful approach to enhancing decision-making and performance monitoring. By collecting and analyzing data, visualizing it through BI tools, and extracting actionable insights, businesses can support their decision-making processes with accurate and real-time information. The proposed strategy leverages key insights from relevant studies to create a comprehensive framework for implementing BI in a business context.

4.4 Implementing decision intelligence for enhanced predictive analytics and scenario planning

Objective: To enhance decision-making processes by integrating Decision Intelligence principles, utilizing advanced analytics and machine learning to provide deeper insights and predictive capabilities.

Overview: Decision Intelligence enhances decision-making by applying advanced analytical models and machine learning algorithms to make predictions and forecasts. Conducting scenario analyses to evaluate different possibilities and outcomes optimizes decision-making by providing actionable insights. The following strategy outlines how Decision Intelligence can be effectively integrated into business practices.

Proposed strategy:

1. Implement Predictive Models to Forecast Sales and Inventory Needs:

    - *Objective:* Use advanced analytics to accurately forecast sales and inventory requirements.

    - *Action:* Utilize insights from Mariani et al.'s study on business intelligence and big data in the hospitality and tourism industry to develop predictive models [82].

    - *Benefit:* Enables accurate sales forecasts and optimal inventory management, reducing costs and improving efficiency.

2. Conduct Scenario Analyses to Plan for Different Market Conditions:

    - *Objective:* Evaluate various market scenarios to prepare for different market conditions.

    - *Action:* Apply methodologies from Phillips-Wren et al.'s research on business analytics in the context of big data to conduct scenario analyses [83].

    - *Benefit:* Enhances strategic planning by providing a comprehensive understanding of potential outcomes, enabling proactive decision-making.

3. Integrate Ethical Considerations in Predictive Analytics:

    - *Objective:* Ensure ethical decision-making in the application of predictive analytics.

    - *Action:* Incorporate ethical guidelines from Beil et al.'s research on ethical considerations about artificial intelligence for prognostication [84].

    - *Benefit:* Promotes responsible use of AI and machine learning in decision-making processes, ensuring fairness and transparency.

4. Adopt Novel Approaches in Business Intelligence for Big Data Analytics:

    - *Objective:* Leverage cutting-edge techniques for big data analytics.

    - *Action:* Implement Mishra et al.'s novel approach in business intelligence using unsupervised techniques for big data analytics [29].

    - *Benefit:* Provides advanced analytical capabilities, enabling deeper insights and improved decision-making.

Integrating Decision Intelligence into business practices offers a powerful approach to enhancing predictive analytics and scenario planning. By implementing predictive models, conducting scenario analyses, considering ethical implications,

DOI: http://dx.doi.org/10.5772/intechopen.1007794

and adopting novel big data analytics techniques, businesses can optimize their decision-making processes. The proposed strategy leverages key insights from relevant studies to create a comprehensive framework for implementing Decision Intelligence in a business context.

4.5 Implementing strategic intelligence for effective long-term strategy formulation and execution

Objective: To formulate and execute effective long-term strategies by integrating Strategic Intelligence principles, utilizing insights from System Thinking, Game Theory, Business Intelligence, and Decision Intelligence.

Overview: Strategic Intelligence provides the framework for setting long-term goals, analyzing the competitive landscape, and identifying future trends. By integrating insights from various disciplines, businesses can create comprehensive strategies that align with their long-term objectives. The following strategy outlines how Strategic Intelligence can be effectively integrated into business practices.

Proposed strategy:

1. Set Long-Term Growth Targets and Expansion Plans:

    o *Objective:* Establish clear long-term goals and detailed expansion plans.

    o *Action:* Utilize insights from Ahmadi et al.'s model of manager's strategic intelligence to set strategic goals and develop actionable plans [85].

    o *Benefit:* Ensures alignment of long-term objectives with organizational capabilities and market opportunities.

2. Analyze Market Trends and Competitive Landscape:

    o *Objective:* Identify strategic opportunities and threats by analyzing market trends and the competitive environment.

    o *Action:* Apply methodologies from Pellissier and Kruger's study on strategic intelligence as a strategic management tool to conduct thorough market and competitive analyses [58].

    o Enhances the ability to anticipate market changes and adapt strategies accordingly.

3. Integrate Insights from Business Intelligence and Advanced Analytics:

    o *Objective:* Leverage data and analytics to support strategic decision-making.

    o *Action:* Implement Khaddam's approach to using business intelligence tools and organizational ambidexterity to enhance strategic planning processes [57].

    o *Benefit:* Provides deeper insights into business performance and market conditions, supporting informed strategic decisions.

4. Utilize Business Intelligence Systems for Strategic Decision-Making:

   - *Objective:* Combine operational data with analytical tools to support strategic intelligence.

   - *Action:* Follow the guidelines from Puklavec et al.'s study on the determinants of business intelligence system adoption stages to effectively utilize BI systems for strategic purposes [86].

   - *Benefit:* Enables the presentation of complex and competitive information to decision-makers, enhancing strategic intelligence.

5. Integrate System Thinking and Game Theory for Comprehensive Strategies:

   - *Objective:* Develop holistic and adaptive strategies by integrating multiple disciplines.

   - *Action:* Incorporate insights from Negash's research on business intelligence systems and their role in presenting complex information to support strategic intelligence [87].

   - *Benefit:* Creates comprehensive strategies that consider system-wide interactions and competitive dynamics.

Integrating Strategic Intelligence into business practices offers a robust approach to formulating and executing effective long-term strategies. By setting clear growth targets, analyzing market trends, leveraging advanced analytics, and integrating insights from various disciplines, businesses can enhance their strategic decision-making processes. The proposed strategy leverages key insights from relevant studies to create a comprehensive framework for implementing Strategic Intelligence in a business context.

4.6 Implementing swarm intelligence for enhanced optimization and decision-making

Objective: To enhance operational efficiency and adaptability by integrating Swarm Intelligence principles, leveraging decentralized and adaptive solutions for optimization and decision-making.

Overview: Swarm Intelligence offers decentralized, adaptive solutions for tasks such as route planning, resource allocation, and network optimization. By using swarm intelligence algorithms, businesses can leverage the collective behavior of simple agents to solve complex problems efficiently. The following strategy outlines how Swarm Intelligence can be effectively integrated into business practices.

Proposed strategy:

1. Optimize Logistics and Supply Chain Operations:

   - *Objective:* Improve logistics and supply chain efficiency using swarm intelligence algorithms.

- *Action:* Implement the hybrid multi-objective particle swarm optimization strategy discussed by Yang et al. to optimize logistics and supply chain operations [88].
- *Benefit:* Enhances the efficiency of logistics operations by optimizing routes and resource allocation.

2. Implement Adaptive Inventory Management Systems:
   - *Objective:* Develop adaptive inventory management systems that respond to changing demand.
   - *Action:* Apply the Particle Swarm Optimization Gray Wolf Fusion Algorithm described by Jiang et al. to create adaptive inventory management systems [89].
   - *Benefit:* Improves inventory management by dynamically adjusting to changes in demand and reducing stockouts and overstock situations.
3. Optimize Joint Ordering Strategies:
   - *Objective:* Optimize joint ordering strategies in supply chain operations.
   - *Action:* Utilize the improved swarm intelligence algorithm for multi-item joint ordering strategies introduced by Huang and Yang [90].
   - *Benefit:* Increases efficiency in supply chain management by optimizing ordering processes for multiple items simultaneously.
4. Enhance Supply Chain Market Analytics Management:
   - *Objective:* Use swarm intelligence techniques to improve supply chain market analytics management.
   - *Action:* Implement the swarm intelligence techniques for supply chain market analytics management explored by Tian et al. [91].
   - *Benefit:* Enhances decision-making processes in supply chain management by providing deeper insights into market trends and demand patterns.
5. Integrate IoT Technology with Swarm Intelligence:
   - *Objective:* Optimize logistics supply chain management systems using advanced technologies.
   - *Action:* Apply the approach from Sun and Gu's study on integrating Internet of Things technology with swarm intelligence for intelligent logistics supply chain management [92].
   - *Benefit:* Improves operational efficiency by leveraging real-time data and adaptive algorithms for logistics management.

Integrating Swarm Intelligence into business practices offers a powerful approach to enhancing optimization and decision-making processes. By implementing swarm intelligence algorithms for logistics, supply chain operations, inventory management, and market analytics, businesses can achieve higher operational efficiency and adaptability. The proposed strategy leverages key insights from relevant studies to create a comprehensive framework for implementing Swarm Intelligence in a business context.

5. Integrating measurement instruments and intelligences: A comprehensive guide

Integrating System Thinking, Game Theory, Business Intelligence, Decision Intelligence, Strategic Intelligence, and Swarm Intelligence into your business processes can be a transformative journey. The following step-by-step guide, combined with the metaphorical use of measurement instruments from various scientific fields, provides a structured approach to enhance decision-making, optimize operations, and achieve strategic goals.

6. Step-by-step guide to integrating intelligences

6.1 Assessment and goal setting

Thermometer

- *Scientific field:* Physics, Meteorology
- *Business application:* Measure employee satisfaction and motivation. Determine the "temperature" of innovation projects, indicating their acceptance and interest levels. Also useful for tracking customer satisfaction and loyalty
- *Relevant concepts:* Business Intelligence, Decision Intelligence, Organizational Behavior, Employee Engagement, Customer Relationship Management

Actions:

- *Assess current state:* Conduct a thorough analysis of your current business processes, decision-making frameworks, and data management systems.
- *Define objectives:* Clearly define what you aim to achieve with the integration, such as improved decision-making, enhanced efficiency, or better strategic planning.

6.2 Building awareness and understanding

Barometer

- *Scientific field: Meteorology*

- *Business application*: Evaluate market conditions, competitive pressures, and economic situations. Monitor the impact of innovation projects on the market. Can also be used to assess financial performance and economic risks
- *Relevant concepts*: Market Analysis, Competitive Intelligence, Financial Risk Management, Strategic Intelligence, Game Theory

Actions:

- *Training and education:* Provide training sessions and workshops to your team to build awareness and understanding of each intelligence.
- *Expert consultation:* Engage with experts in each field to gain insights and guidance on best practices for integration.

6.3 System thinking

Compass

- *Scientific field:* Geography, Navigation
- *Business application:* Provide guidance in strategic planning and decision-making processes. Ensure innovation projects are on the right strategic path. Useful for general strategic planning and goal setting.
- *Relevant concepts:* System Thinking, Strategic Planning, Corporate Governance, Vision and Mission Alignment, Strategic Intelligence

Actions:

- *Map processes:* Create detailed maps of your business processes, identifying key components, interdependencies, and feedback loops.
- *Identify leverage points:* Determine areas where small changes can lead to significant improvements.
- *Implement changes:* Make adjustments based on your system maps to improve efficiency and effectiveness.

6.4 Game theory

Hygrometer

- *Scientific field:* Meteorology, Environmental Sciences
- *Business application:* Assess the workplace atmosphere. Measure the spread and adoption level of innovation projects within the organization. Can also be used for evaluating internal communication and employee engagement.
- *Relevant concepts:* Organizational Culture, Internal Communications, Change Management, Swarm Intelligence

Actions:

- *Strategic analysis:* Analyze the strategic interactions within your business environment, including competitors, partners, and market conditions.
- *Scenario planning:* Develop various scenarios and strategies based on potential actions of competitors and market changes.
- *Decision framework:* Establish a decision framework that incorporates game theory principles for strategic planning and competitive strategy.

6.5 Business intelligence

Anemometer

- *Scientific field*: Meteorology
- *Business application*: Measure the speed of innovation. Evaluate the pace and effectiveness of R&D processes. Also useful for monitoring the efficiency and speed of business processes.
- *Relevant concepts*: Innovation Management, Process Optimization, Agile Methodologies, Decision Intelligence

Actions:

- *Data collection:* Set up systems to collect data on key performance indicators, market trends, and customer behavior.
- *BI tools:* Implement BI tools to integrate, analyze, and visualize data.
- *Regular reports:* Generate regular reports and dashboards to provide actionable insights for decision-makers.

6.6 Decision intelligence

Altimeter

- *Scientific field:* Aviation, Mountaineering
- *Business application:* Measure the "height" of business success, i.e., growth and progress levels. Evaluate the impact of innovation projects on the company. Useful for sales performance and market share analysis.
- *Relevant concepts:* Growth Strategy, Performance Metrics, Market Penetration, Business Intelligence, Strategic Intelligence

Actions:

- *Advanced analytics:* Use advanced analytical models and machine learning algorithms to enhance predictive capabilities.

- *Scenario analysis:* Conduct scenario analyses to evaluate different possibilities and outcomes.

- *Decision support systems:* Implement decision support systems that leverage these advanced analytics to inform and optimize decision-making processes.

6.7 Strategic intelligence

Seismometer

- *Scientific field*: Geology

- *Business application*: Monitor significant changes or crises within the business. Measure the major transformations and impacts of innovation projects. Useful in managing major organizational changes and crisis management processes.

- *Relevant concepts*: Crisis Management, Change Management, Risk Assessment, System Thinking

Actions:

- *Long-term planning:* Set long-term strategic goals and develop detailed plans to achieve them.

- *Competitive analysis:* Continuously analyze the competitive landscape and future trends.

- *Integrated insights:* Integrate insights from system thinking, game theory, and business intelligence to refine strategies.

6.8 Swarm intelligence

Spectrometer

- *Scientific field:* Chemistry, Physics

- *Business application:* Analyze various performance metrics of innovation projects. Provide comprehensive evaluations of R&D activities. Useful for marketing analysis and customer segmentation.

- *Relevant concepts:* Data Analytics, Market Segmentation, Product Development, Business Intelligence

Actions:

- *Decentralized algorithms:* Implement swarm intelligence algorithms for tasks like route planning, resource allocation, and network optimization. These algorithms can be particularly powerful when combined with Decision Intelligence, enabling real-time, data-driven adjustments in rapidly changing environments.

- *Adaptive systems:* Develop adaptive inventory management systems that respond to changing demand.
- *Collaboration tools:* Utilize tools that enable decentralized decision-making and collective problem-solving.

6.9 Implementation and monitoring

Flow Meter

- *Scientific field:* Engineering, Hydrodynamics
- *Business application:* Evaluate the efficiency and flow of business processes. Measure the efficiency of innovation ideas from inception to market launch. Useful for monitoring supply chain and logistics performance.
- *Relevant concepts:* Supply Chain Management, Logistics Optimization, Lean Management, System Thinking

Actions:

- *Pilot programs:* Start with pilot programs to test the integration of these intelligences in specific areas of your business.
- *Monitor progress:* Regularly monitor and evaluate the effectiveness of the integrated systems.
- *Continuous improvement:* Make adjustments and improvements based on feedback and performance metrics.

6.10 Scaling and integration

Multimeter

- *Scientific field:* Electrical Engineering
- *Business application:* Measure various performance metrics with a single tool. Evaluate the overall performance of innovation and R&D projects. Useful for financial analysis and budget evaluations.
- *Relevant concepts:* Performance Management, Financial Analysis, Balanced Scorecard, Business Intelligence

Actions:

- *Expand integration:* Gradually scale the integration to other areas of the business.
- *Full integration:* Aim for a fully integrated system where all intelligence works seamlessly together to support business operations and strategic planning.

DOI: http://dx.doi.org/10.5772/intechopen.1007794

7. Conclusion

Starting the integration of these intelligences requires careful planning, education, and incremental implementation. By following these steps, businesses can harness the power of System Thinking, Game Theory, Business Intelligence, Decision Intelligence, Strategic Intelligence, and Swarm Intelligence to enhance decision-making, optimize operations, and achieve strategic goals.

Incorporating measurement instruments from various scientific fields into business sciences allows companies to develop more effective and efficient management strategies in complex and dynamic business environments. Measurement instruments inspired by fields such as physics, chemistry, engineering, and environmental sciences can be used to evaluate and optimize the performance of business functions. Integrating concepts like System Thinking, Game Theory, Business Intelligence, Decision Intelligence, Strategic Intelligence, and Swarm Intelligence further enhances the capacity for holistic and informed decision-making. Below are examples of how these measurement tools can be adapted to business applications:

1. Thermometer:
   - *Scientific field:* Physics, Meteorology
   - *Business application:* Measure employee satisfaction and motivation. Determine the "temperature" of innovation projects, indicating their acceptance and interest levels. Also useful for tracking customer satisfaction and loyalty.
   - *Relevant concepts:* Business Intelligence, Decision Intelligence, Organizational Behavior, Employee Engagement, Customer Relationship Management
2. Barometer:
   - *Scientific field:* Meteorology
   - *Business application:* Evaluate market conditions, competitive pressures, and economic situations. Monitor the impact of innovation projects on the market. Can also be used to assess financial performance and economic risks.
   - *Relevant concepts:* Market Analysis, Competitive Intelligence, Financial Risk Management, Strategic Intelligence, Game Theory
3. Compass:
   - *Scientific field:* Geography, Navigation
   - *Business application:* Provide guidance in strategic planning and decision-making processes. Ensure innovation projects are on the right strategic path. Useful for general strategic planning and goal setting.
   - *Relevant concepts:* System Thinking, Strategic Planning, Corporate Governance, Vision and Mission Alignment, Strategic Intelligence

4. Hygrometer:

    - *Scientific field:* Meteorology, Environmental Sciences
    - *Business application:* Assess the workplace atmosphere. Measure the spread and adoption level of innovation projects within the organization. Can also be used for evaluating internal communication and employee engagement.
    - *Relevant concepts:* Organizational Culture, Internal Communications, Change Management, Swarm Intelligence

5. Anemometer:

    - *Scientific field:* Meteorology
    - *Business application:* Measure the speed of innovation. Evaluate the pace and effectiveness of R&D processes. Also useful for monitoring the efficiency and speed of business processes.
    - *Relevant concepts:* Innovation Management, Process Optimization, Agile Methodologies, Decision Intelligence

6. Altimeter:

    - *Scientific field:* Aviation, Mountaineering
    - *Business application:* Measure the "height" of business success, i.e., growth and progress levels. Evaluate the impact of innovation projects on the company. Useful for sales performance and market share analysis.
    - *Relevant concepts:* Growth Strategy, Performance Metrics, Market Penetration, Business Intelligence, Strategic Intelligence

7. Seismometer:

    - *Scientific field:* Geology
    - *Business application:* Monitor significant changes or crises within the business. Measure the major transformations and impacts of innovation projects. Useful in managing major organizational changes and crisis management processes.
    - *Relevant concepts:* Crisis Management, Change Management, Risk Assessment, System Thinking

8. Spectrometer:

    - *Scientific field:* Chemistry, Physics
    - *Business application:* Analyze various performance metrics of innovation projects. Provide comprehensive evaluations of R&D activities. Useful for marketing analysis and customer segmentation.

DOI: http://dx.doi.org/10.5772/intechopen.1007794

- *Relevant concepts:* Data Analytics, Market Segmentation, Product Development, Business Intelligence

9. Flow Meter:

- *Scientific field:* Engineering, Hydrodynamics
- *Business application:* Evaluate the efficiency and flow of business processes. Measure the efficiency of innovation ideas from inception to market launch. Useful for monitoring supply chain and logistics performance.
- *Relevant concepts:* Supply Chain Management, Logistics Optimization, Lean Management, System Thinking

10. Multimeter:

- *Scientific field:* Electrical Engineering
- *Business application:* Measure various performance metrics with a single tool. Evaluate the overall performance of innovation and R&D projects. Useful for financial analysis and budget evaluations.
- *Relevant concepts:* Performance Management, Financial Analysis, Balanced Scorecard, Business Intelligence

11. Oscilloscope:

- *Scientific field:* Electronics
- *Business application:* Analyze changes and fluctuations over time. Monitor performance changes from the beginning to the completion of innovation projects. Useful for evaluating the impact of marketing campaigns and customer feedback.
- *Relevant concepts:* Marketing Analytics, Customer Feedback Analysis, Performance Tracking, Decision Intelligence

12. Manometer:

- *Scientific field:* Physics, Engineering
- *Business application:* Measure pressure levels within the business. Evaluate the stress and challenges in innovation projects. Useful for managing employee stress and workload.
- *Relevant concepts:* Stress Management, Workload Management, Employee Wellbeing, Swarm Intelligence

13. Accelerometer:

    - *Scientific field:* Physics, Engineering
    - *Business application:* Measure the speed and acceleration of innovation projects. Evaluate the development and market launch speed of new technologies. Useful for monitoring production speed and operational efficiency.
    - *Relevant concepts:* Time-to-Market, Operational Efficiency, Productivity Analysis, Decision Intelligence

14. Calorimeter:

    - *Scientific field:* Chemistry, Physics
    - *Business application:* Evaluate the energy and resource use of innovation projects. Measure the resource efficiency and sustainability of R&D projects. Useful for assessing overall energy efficiency and cost management.
    - *Relevant concepts:* Resource Management, Sustainability, Cost Efficiency, System Thinking

15. Dynamometer:

    - *Scientific field:* Mechanical Engineering
    - *Business application:* Measure the strength and impact of innovation projects. Evaluate the impact of R&D activities on the company's competitive strength and market position. Useful for assessing employee performance and productivity.
    - *Relevant concepts:* Competitive Advantage, Employee Performance, Productivity Metrics, Strategic Intelligence

By drawing inspiration from these measurement instruments, creative and effective methods can be developed to evaluate performance, efficiency, strategy, and impact across various business functions (e.g., marketing, finance, human resources, and operations). These approaches can enhance a company's overall management capacity and competitive strength. The integration of big data analytics allows for more accurate and comprehensive use of these measurement and evaluation tools, providing businesses with deep insights and real-time data. Thus, an interdisciplinary approach combined with big data technologies enriches measurement and evaluation processes, enabling businesses to adapt to changing market conditions and gain strategic advantages. Big data expands companies' measurement and evaluation capacities, making these tools more effective and allowing for deeper and more strategic decision-making in business management.

In the era of big data, businesses can significantly enhance their measurement capabilities by integrating tools and methodologies from various scientific fields.

DOI: http://dx.doi.org/10.5772/intechopen.1007794

Emphasizing the principle that "you cannot manage what you cannot measure," companies can adopt instruments from physics, chemistry, engineering, and environmental sciences to develop more precise and innovative methods for assessing performance, optimizing processes, and making informed strategic decisions. Physics is not a distant field for businesses; new concepts and measurement tools relevant to business can be developed from each scientific discipline. For example, thermometers and barometers commonly used in environmental sciences can be adapted to measure employee satisfaction or market pressures, thereby providing critical data for human resource management and strategic planning.

Fractal geometry also offers valuable insights for businesses; self-similar patterns in fractals can be used to model market trends or customer behaviors, revealing complexities that traditional analysis might overlook. The alignment of data and patterns functions similarly to blood type matching and blood donation processes; in the age of artificial intelligence, such efforts will accelerate. Those who detect the unseen and apply it to their businesses will gain a competitive advantage.

By incorporating these scientific tools into business intelligence systems, organizations can leverage diverse data sources for deeper insights and support long-term growth. Therefore, expanding the horizons of business measurement through interdisciplinary approaches allows companies to enhance their strategic capabilities and succeed in competitive environments.

Author details

Eyup Akcetin
Department of Accounting and Financial Management, Seydikemer School of Applied Sciences, Mugla Sitki Kocman University, Mugla, Turkey

*Address all correspondence to: eyup.akcetin@mu.edu.tr

References

[1] Richmond BJ. Systems thinking: Critical thinking skills for the 1990s and beyond. System Dynamics Review. 1993;**9**(2):113-133. DOI: 10.1002/sdr.4260090203

[2] Orgill M, York S, MacKellar J. Introduction to systems thinking for the chemistry education community. Journal of Chemical Education. 2019;**96**(12):2720-2729. DOI: 10.1021/acs.jchemed.9b00169

[3] Kwamie A, Ha S, Ghaffar A. Applied systems thinking: Unlocking theory, evidence and practice for health policy and systems research. Health Policy and Planning. 2021;**36**(10):1715-1717. DOI: 10.1093/heapol/czab062

[4] Haynes A, Garvey K, Davidson S, Milat A. What can policy-makers get out of systems thinking? Policy partners' experiences of a systems-focused research collaboration in preventive health. International Journal of Health Policy and Management. 2019;**9**(2):65-76. DOI: 10.15171/ijhpm.2019.86

[5] Flynn AB et al. Future directions for systems thinking in chemistry education: Putting the pieces together. Journal of Chemical Education. 2019;**96**(12):3000-3005. DOI: 10.1021/acs.jchemed.9b00637

[6] McNab D, McKay J, Shorrock S, Luty S, Bowie P. Development and application of 'systems thinking' principles for quality improvement. BMJ Open Quality. 2020;**9**(1):e000714. DOI: 10.1136/bmjoq-2019-000714

[7] Bui H-T, Galanou E. Translation of systems thinking to organizational goals: A systematic review. Journal of General Management. 2022;**47**(4):233-245. DOI: 10.1177/03063070211035749

[8] Skaržauskienė A. Managing complexity: Systems thinking as a catalyst of the organization performance. Measuring Business Excellence. 2010;**14**(4):49-64. DOI: 10.1108/13683041011093758

[9] Khan S. Overcoming barriers to applying systems thinking mental models in policy-making comment on "what can policy-makers get out of systems thinking? Policy partners' experiences of a systems-focused research collaboration in preventive health". International Journal of Health Policy and Management. 15 Mar 2021;**10**(5):281-283. DOI: 10.34172/ijhpm.2020.50. PMID: 32610795; PMCID: PMC9056185

[10] Samuelson L. Game theory in economics and beyond. Journal of Economic Perspectives. 2016;**30**(4):107-130. DOI: 10.1257/jep.30.4.107

[11] Wu J. The state-of-art applications of game theory in computer science. Highlights in Science Engineering and Technology. 2023;**49**:280-286. DOI: 10.54097/hset.v49i.8519

[12] Xu M. Some fundamental model from game theory. Journal of Education Humanities and Social Sciences. 2022;**2**:99-104. DOI: 10.54097/ehss.v2i.768

[13] Lieberman EJ, Hauert C, Nowak MA. Evolutionary dynamics on graphs. Nature. 2005;**433**(7023):312-316. DOI: 10.1038/nature03204

[14] Shakarian P, Roos P, Johnson AN. A review of evolutionary graph theory with applications to game theory. Biosystems. 2012;**107**(2):66-80. DOI: 10.1016/j.biosystems.2011.09.006

[15] Morini L. Playful participatory mapping: Co-creating games to

foster systems thinking. European Conference on Games Based Learning. 2022;**16**(1):387-395. DOI: 10.34190/ecgbl.16.1.591

[16] Storey BD, Butler J. Complexity thinking in PE: Game-centred approaches, games as complex adaptive systems, and ecological values. Physical Education and Sport Pedagogy. 2012;**18**(2):133-149. DOI: 10.1080/17408989.2011.649721

[17] Benedetto AD. Modeling rationality in marketing decision-making with game theory. Journal of the Academy of Marketing Science. 1987;**15**(4):22-32. DOI: 10.1177/009207038701500404

[18] Blumentritt T, Mathews T, Marchisio G. Game theory and family business succession. Family Business Review. 2012;**26**(1):51-67. DOI: 10.1177/0894486512447811

[19] Su Y, Backlund P, Engström H. Business intelligence challenges for independent game publishing. International Journal of Computer Games Technology. 2020;**2020**:1-8. DOI: 10.1155/2020/5395187

[20] Bauckhage C, Kersting K, Sifa R, Thurau C, Drachen A, Canossa A. How Players Lose Interest in Playing a Game: An Empirical Study Based on Distributions of Total Playing Times. 2012. doi:10.1109/cig.2012.6374148

[21] Zhou L, Liu W. An approach for community detection in social networks based on cooperative games theory. Expert Systems. 2015;**33**(2):176-188. DOI: 10.1111/exsy.12141

[22] Esmaeilyfard R, Salehi A. An incentive mechanism for integration of business applications between organizations. RAIRO - Operations Research. 2023;**57**(2):427-446. DOI: 10.1051/ro/2023024

[23] Wang S, Wu J, Zhang Y. Consumer preference–enabled intelligent energy management for smart cities using game theoretic social tie. International Journal of Distributed Sensor Networks. 2018;**14**(4):155014771877323. DOI: 10.1177/1550147718773235

[24] Abdalzaher MS, Seddik KG, Elsabrouty M, Muta O, Furukawa H, Abdel-Rahman AB. Game theory meets wireless sensor networks security requirements and threats mitigation: A survey. Sensors. 2016;**16**(7):1003. DOI: 10.3390/s16071003

[25] Muntean M. Business intelligence issues for sustainability projects. Sustainability. 2018;**10**(2):335. DOI: 10.3390/su10020335

[26] Kapo A, Turulja L, Zaimović T, Mehić S. Examining the effect of user satisfaction and business intelligence system usage on individual job performance. Management. 2021;**26**(2):43-62. DOI: 10.30924/mjcmi.26.2.3

[27] Rouhani S, Ravasan AZ, Ashrafi A, Afshari S. Business intelligence systems adoption model. Journal of Organizational and End User Computing. 2018;**30**(2):43-70. DOI: 10.4018/joeuc.2018040103

[28] Hamad F, Al-Aamr R, Jabbar SA, Fakhouri HN. Business intelligence in academic libraries in Jordan: Opportunities and challenges. IFLA Journal. 2020;**47**(1):37-50. DOI: 10.1177/0340035220931882

[29] Mishra D, Johari K, Ghildiyal S, Upadhyay AK, Sharma S. A novel approach in business intelligence for big data analytics using an unsupervised technique. ECS Transactions. 2022;**107**(1):12525-12533. DOI: 10.1149/10701.12525ecst

[30] Fabiyi O, Olanipekun WD. Evaluating the Impact of Critical Success Factors of Business Intelligence on Firm Performance. 2021 DOI:10.22624/aims/isteams-2021/v28n3p10

[31] Calof J, Richards G, Santilli P. Integration of business intelligence with corporate strategic management. Journal of Intelligence Studies in Business. 2017;7(3). DOI: 10.37380/jisib.v7i3.280

[32] Tank D. Enable better and timelier decision-making using real-time business intelligence system. International Journal of Information Engineering and Electronic Business. 2015;7(1):43-48. DOI: 10.5815/ijieeb.2015.01.06

[33] Rothschild R. Ten simple lessons in strategy from the games firms play. Management Decision. 1995;**33**(9):24-29. DOI: 10.1108/00251749510098955

[34] Fan W. A comprehensive analysis of game theory on multi-agent reinforcement. Highlights in Science Engineering and Technology. 2024;**85**:77-88. DOI: 10.54097/gv6fpz53

[35] Chithra S, Vinoba V. The study of strategic and extensive form of non cooperative game theory. International Journal of Mathematics Trends and Technology. 2015;**22**(1):25-31. DOI: 10.14445/22315373/ijmtt-v22p505

[36] Naumov P, Yuan Y. Intelligence in strategic games. Journal of Artificial Intelligence Research. 2021;**71**:521-556. DOI: 10.1613/jair.1.12883

[37] Agyepong IA. Wood already touched by fire is not hard to set alight' comment on 'constraints to applying systems thinking concepts in health systems: A regional perspective from surveying stakeholders in eastern Mediterranean countries. International Journal of Health Policy and Management. 2015;**4**(3):191-193. DOI: 10.15171/ijhpm.2015.34

[38] Chan-Lau JA. Regulatory capital charges for too-connected-to-fail institutions: A practical proposal. Financial Markets Institutions and Instruments. 2010;**19**(5):355-379. DOI: 10.1111/j.1468-0416.2010.00161.x

[39] Cortes F, Lindner P, Malik S, Basurto MAS. A comprehensive multi-sector tool for analysis of systemic risk and interconnectedness (SyRIN). IMF Working Paper. 2018;**18**(14):1. DOI: 10.5089/9781484338605.001

[40] Wächter P. Thinking in systems—A primer. Environmental Politics. 2011;**20**(4):595-596. DOI: 10.1080/09644016.2011.589585

[41] Pearson DE, Clark TJ, Hahn PG. Evaluating unintended consequences of intentional species introductions and eradications for improved conservation management. Conservation Biology. Feb 2022;**36**(1):e13734. DOI: 10.1111/cobi.13734. Epub 2021 May 31. PMID: 33734489; PMCID: PMC9291768

[42] Harst MVD, Koch D, Brink MCLVD. A review of the unintended gender effects of international development efforts. Public Administration and Development. 2023;**43**(4):280-292. DOI: 10.1002/pad.2023

[43] Harris LC, Ogbonna EO. The unintended consequences of culture interventions: A study of unexpected outcomes. British Journal of Management. 2002;**13**(1):31-49. DOI: 10.1111/1467-8551.00221

[44] Persson T, Roland G, Tabellini G. Comparative politics and public finance. Journal of Political Economy. 2000;**108**(6):1121-1161. DOI: 10.1086/317686

[45] Dellal A, Owen A, Wong DP, Krustrup P, Exsel MV, Mallo J. Technical and physical demands of small vs. large sided games in relation to playing position in elite soccer. Human Movement Science. 2012;**31**(4):957-969. DOI: 10.1016/j.humov.2011.08.013

[46] Zietsma C, Lawrence TB. Institutional work in the transformation of an organizational field: The interplay of boundary work and practice work. Administrative Science Quarterly. 2010;**55**(2):189-221. DOI: 10.2189/asqu.2010.55.2.189

[47] Mørk BE, Hoholm T, Maaninen-Olsson E, Aanestad M. Changing practice through boundary organizing: A case from medical R& D. Human Relations. 2012;**65**(2):263-288. DOI: 10.1177/0018726711429192

[48] Wang Q, Gao S. Network architecture and firm performance: A resources-based view. Frontiers of Business Research in China. 2011;**5**(4):559-579. DOI: 10.1007/s11782-011-0145-0

[49] Huang R, Fang C, Zhang B, Tang Y. Transformations of phosphorus speciation during (hydro)thermal treatments of animal manures. Environmental Science and Technology. 2018;**52**(5):3016-3026. DOI: 10.1021/acs.est.7b05203

[50] Gupta HC, Barua MK. Modelling cause and effect relationship among enablers of innovation in SMEs. Benchmarking an International Journal. 2018;**25**(5):1597-1622. DOI: 10.1108/bij-03-2017-0050

[51] Pescaroli G, Nones M. Cascading events, technology and the floods directive: Future challenges. E3s Web of Conferences. 2016;7:07003. DOI: 10.1051/e3sconf/20160707003

[52] Rycroft-Malone J. From linear to complicated to complex comment on 'using complexity and network concepts to inform healthcare knowledge translation. International Journal of Health Policy and Management. 2018;**7**(6):566-568. DOI: 10.15171/ijhpm.2018.02

[53] Jirout J, Klahr D. Children's scientific curiosity: In search of an operational definition of an elusive concept. Developmental Review. 2012;**32**(2):125-160. DOI: 10.1016/j.dr.2012.04.002

[54] Sun D, Hyland P, Cui H. A designed framework for delivering systems thinking skills to small business managers. Systems. 2014;**2**(3):297-312. DOI: 10.3390/systems2030297

[55] Friedman D. On economic applications of evolutionary game theory. Journal of Evolutionary Economics. 1998;**8**(1):15-43. DOI: 10.1007/s001910050054

[56] Rausch P. Business Intelligence and Performance Management. 2013. DOI:10.1007/978-1-4471-4866-1

[57] Khaddam A. Enhancing strategic decision-making: The role of business intelligence tools and organizational ambidexterity. Problems and Perspectives in Management. 2024;**22**(1):716-727. DOI: 10.21511/ppm.22(1).2024.56

[58] Pellissier R, Kruger J-P. A study of strategic intelligence as a strategic management tool in the long-term insurance industry in South Africa. European Business Review. 2011;**23**(6):609-631. DOI: 10.1108/09555341111175435

[59] Zhou Y, Rao BG, Wang W. UAV swarm intelligence: Recent advances and future trends. IEEE Access.

2020;**8**:183856-183878. DOI: 10.1109/access.2020.3028865

[60] Mu Y, Li B, An D, Wei Y. Three-dimensional route planning based on the beetle swarm optimization algorithm. IEEE Access. 2019;**7**:117804-117813. DOI: 10.1109/access.2019.2935835

[61] Zangana HM. Unveiling the collective wisdom: A review of swarm intelligence in problem solving and optimization. Inform Jurnal Ilmiah Bidang Teknologi Informasi Dan Komunikasi. 2024;**9**(2):101-110. DOI: 10.25139/inform.v9i2.7934

[62] Yejiao W. RFID network planning of smart factory based on swarm intelligent optimization algorithm: A review. IEEE Access. 2024;**12**:64980-64996. DOI: 10.1109/access.2024.3397402

[63] Oyekunle DOT. Swarm Intelligence and Decision-Making: The Role of Swarm Intelligence in Decision. 2024. DOI:10.2139/ssrn.4849581

[64] Yang X-S. Firefly Algorithms for Multimodal Optimization. 2009. pp. 169-178. DOI:10.1007/978-3-642-04944-6_14

[65] Brezočnik L, Fister I, Podgorelec V. Swarm intelligence algorithms for feature selection: A review. Applied Sciences. 2018;**8**(9):1521. DOI: 10.3390/app8091521

[66] Marshall JAR, Bogacz R, Dornhaus A, Planquè R, Kovacs T, Franks NR. On optimal decision-making in brains and social insect colonies. Journal of The Royal Society Interface. 2009;**6**(40):1065-1074. DOI: 10.1098/rsif.2008.0511

[67] Couzin ID. Collective cognition in animal groups. Trends in Cognitive Sciences. 2009;**13**(1):36-43. DOI: 10.1016/j.tics.2008.10.002

[68] Metcalf LE, Askay DA, Rosenberg L. Keeping humans in the loop: Pooling knowledge through artificial swarm intelligence to improve business decision making. California Management Review. 2019;**61**(4):84-109. DOI: 10.1177/0008125619862256

[69] Sweeney LB, Sterman JD. Bathtub dynamics: Initial results of a systems thinking inventory. System Dynamics Review. 2000;**16**(4):249-286. DOI: 10.1002/sdr.198

[70] Beamon BM. Measuring supply chain performance. International Journal of Operations and Production Management. 1999;**19**(3):275-292. DOI: 10.1108/01443579910249714

[71] Owen B et al. Understanding a successful obesity prevention initiative in children under 5 from a systems perspective. PLoS One. 2018;**13**(3):e0195141. DOI: 10.1371/journal.pone.0195141

[72] Kapsali M. Systems thinking in innovation project management: A match that works. International Journal of Project Management. 2011;**29**(4):396-407. DOI: 10.1016/j.ijproman.2011.01.003

[73] Wilden D, Sadler I. The prevalence of systems thinking in supply chain management: A systematic literature review. Systemic Practice and Action Research. 2021;**35**(4):491-526. DOI: 10.1007/s11213-021-09578-5

[74] Meyer D. Quantum strategies. Physical Review Letters. 1999;**82**(5):1052-1055. DOI: 10.1103/physrevlett.82.1052

[75] Ho E, Rajagopalan A, Skvortsov A, Arulampalam S, Piraveenan M. Game theory in defence applications: A review. Sensors. 2022;**22**(3):1032. DOI: 10.3390/s22031032

[76] Abedian M, Amindoust A, Maddahi R, Jouzdani J. A game theory approach to selecting marketing-mix strategies. Journal of Advances in Management Research. 2021;**19**(1):139-158. DOI: 10.1108/jamr-10-2020-0264

[77] Nie P, Matsuhisa T, Wang XH, Zhang P. Game theory and applications in economics. Journal of Applied Mathematics. 2014;**2014**:1-2. DOI: 10.1155/2014/936192

[78] Christanto HJ. Game theory analysis on marketing strategy determination of KAI access and Traveloka based on usability of HCI (human-computer interaction). Journal of Information Systems and Informatics. 2022;**4**(3):665-672. DOI: 10.51519/journalisi.v4i3.300

[79] Feng T, Cummings L, Tweedie D. Exploring integrated thinking in integrated reporting–An exploratory study in Australia. Journal of Intellectual Capital. 2017;**18**(2):330-353. DOI: 10.1108/jic-06-2016-0068

[80] Guthrie J, Rossi FM, Orelli RL. Integrated reporting and integrated thinking in Italian public sector organisations. Meditari Accountancy Research. 2017;**25**(4):553-573. DOI: 10.1108/medar-06-2017-0155

[81] Al-Htaybat K, Alberti-Alhtaybat LV. Integrated thinking leading to integrated reporting: Case study insights from a global player. Accounting Auditing and Accountability Journal. 2018;**31**(5):1435-1460. DOI: 10.1108/aaaj-08-2016-2680

[82] Mariani MM, Baggio R, Fuchs M, Höpken W. Business intelligence and big data in hospitality and tourism: A systematic literature review. International Journal of Contemporary Hospitality Management. 2018;**30**(12):3514-3554. DOI: 10.1108/ijchm-07-2017-0461

[83] Phillips-Wren G, Iyer LS, Kulkarni U, Ariyachandra T. Business analytics in the context of big data: A roadmap for research. Communications of the Association for Information Systems. 2015;**37**. DOI: 10.17705/1CAIS.03723

[84] Beil M, Proft I, van Heerden D, et al. Ethical considerations about artificial intelligence for prognostication in intensive care. Intensive Care Medicine Experimental. 2019;**7**:70. DOI: 10.1186/s40635-019-0286-6

[85] Ahmadi M, Baei F, Hosseini-Amiri S-M, Moarefi A, Suifan T, Sweis RJ. Proposing a model of manager's strategic intelligence, organization development, and entrepreneurial behavior in organizations. The Journal of Management Development. 2020;**39**(4):559-579. DOI: 10.1108/jmd-11-2018-0317

[86] Puklavec B, Oliveira T, Popovič A. Understanding the determinants of business intelligence system adoption stages. Industrial Management and Data Systems. 2018;**118**(1):236-261. DOI: 10.1108/imds-05-2017-0170

[87] Negash S. Business intelligence. Communications of the Association for Information Systems. 2004;**13**. DOI: 10.17705/1CAIS.01315

[88] Yang M, Liu Y, Yang J. A hybrid multi-objective particle swarm optimization with central control strategy. Computational Intelligence and Neuroscience. 2022;**2022**:1-23. DOI: 10.1155/2022/1522096

[89] Jiang L, Liu K, Wu L. Particle swarm optimization gray wolf fusion algorithm for solving the optimal solution of function. Journal of Physics Conference Series. 2022;**2216**(1):012066. DOI: 10.1088/1742-6596/2216/1/012066

[90] Huang L, Yang J. An improved swarm intelligence algorithm for multi-item joint ordering strategy of cruise ship supply. Mathematical Problems in Engineering. 2020;**2020**:1-9. DOI: 10.1155/2020/5048629

[91] Tian Q, Yin Q, Meng Y. Swarm intelligence technique for supply chain market in logistic analytics management. International Journal of Information Systems and Supply Chain Management. 2022;**15**(4):1-20. DOI: 10.4018/ijisscm.305845

[92] Sun P, Gu L. Optimization of cross-border E-commerce logistics supervision system based on internet of things technology. Complexity. 2021;**2021**:1-11. DOI: 10.1155/2021/4582838